T0351312

Essentials of Modern Spectrum Management

Are you fully up-to-speed on today's modern spectrum management tools? As regulators move away from traditional spectrum management methods, introduce spectrum trading and consider opening up more spectrum to commons, do you understand the implications of these developments for your own networks?

This is the first book to describe and evaluate modern spectrum management tools. Expert authors offer you unique insights into the technical, economic and management issues involved. Auctions, administrative pricing, trading, property rights and spectrum commons are all explained. A series of real-world case studies from around the world is used to highlight the strengths and weaknesses of the various approaches adopted by different regulators, and valuable lessons are drawn from these.

This concise and authoritative resource is a must-have for telecom regulators, network planners, designers and technical managers at mobile and fixed operators and broadcasters, and academics involved in the technology and economics of radio spectrum.

MARTIN CAVE is Professor and Director of the Centre for Management under Regulation at Warwick Business School. He is the author of the *Cave Review* commissioned by the Chancellor into spectrum management in the UK.

CHRIS DOYLE is an Associate Fellow at the Centre for Management under Regulation at Warwick Business School.

WILLIAM WEBB is Head of Research and Development and Senior Technologist at Ofcom, a Visiting Professor at Surrey University and Fellow of the Royal Academy of Engineering.

The Cambridge Wireless Essentials Series

Series Editors

WILLIAM WEBB *Ofcom, UK*
SUDHIR DIXIT *Nokia, US*

A series of concise, practical guides for wireless industry professionals.

Martin Cave, Chris Doyle and William Webb, *Essentials of Modern Spectrum Management*

Forthcoming

Andy Wilton and Tim Charity, *Essentials of Wireless Network Deployment*
Chris Haslett, *Essentials of Radiowave Propagation*
Steve Methley, *Essentials of Wireless Mesh Networking*
Malcolm Macleod and Ian Proudler, *Essentials of Smart Antennas and MIMO*
Stephen Wood and Roberto Aiello, *Essentials of Ultra-Wideband*
David Crawford, *Essentials of Mobile Television*

For further information on any of these titles, the series itself and ordering information see www.cambridge.org/wirelessessentials

Essentials of Modern Spectrum Management

Martin Cave
Warwick Business School

Chris Doyle
Warwick Business School

William Webb
Ofcom, UK

CAMBRIDGE UNIVERSITY PRESS
Cambridge, New York, Melbourne, Madrid, Cape Town,
Singapore, São Paulo, Delhi, Tokyo, Mexico City

Cambridge University Press
The Edinburgh Building, Cambridge CB2 8RU, UK

Published in the United States of America by Cambridge University Press, New York

www.cambridge.org
Information on this title: www.cambridge.org/9780521876698

© Cambridge University Press 2007

This publication is in copyright. Subject to statutory exception
and to the provisions of relevant collective licensing agreements,
no reproduction of any part may take place without the written
permission of Cambridge University Press.

First published 2007
First paperback edition 2011

A catalogue record for this publication is available from the British Library

ISBN 978-0-521-87669-8 Hardback
ISBN 978-0-521-20849-9 Paperback

Cambridge University Press has no responsibility for the persistence or
accuracy of URLs for external or third-party internet websites referred to in
this publication, and does not guarantee that any content on such websites is,
or will remain, accurate or appropriate.

Contents

Acknowledgements

Martin Cave

I am very grateful to a number of people from whom I have learnt about spectrum economics, including William Lehr, Tom Hazlett, Robert Pepper, Evan Kwerel, Michael Goddard, David Hendon and Adele Morris – but they bear no responsibility for the result; I am especially grateful to my co-authors.

Chris Doyle

Over the years I have been fortunate to work with Peter Cramton, Eric van Damme and Paul Milgrom on a number of spectrum assignments and am grateful for their invaluable insights on auctions and spectrum pricing in particular. I am grateful to Phillipa Marks and Brian Williamson of Indepen Economic Consultants, John Burns of Aegis Systems Limited and Charles Chambers formerly of Quotient Associates. I am also grateful to my co-authors for comments. Finally, I should like to thank my wife, Jennifer Smith, who has helped my work in this area.

William Webb

In writing this book I have drawn upon all my experience gained over my years in the industry. I have learnt something from almost everyone I have come into contact with and would thank all of those with whom I have had discussions. Special thanks are due to a number of key individuals. During my time at Multiple Access Communications, Professor Ray Steele, Professor Lajos Hanzo, Dr Ian Wassell and Dr John Williams amongst others have taught me much about the workings of mobile radio systems. At Smith System Engineering (now Detica), Richard Shenton, Dr Glyn Carter and Mike Shannon have provided valuable knowledge as have contacts with a number of others in the industry including Michel Mouly, Mike Watkins, Jim Norton and Phillipa Marks (Indepen). At Motorola I had tremendous guidance from a range of individuals including Sandra Cook, Raghu Rau, John Thode and the immense privilege of discussions with Bob Galvin, ex-CEO. In my work with Institutions I have been privileged to work with John Forrest CBE, Sir David Brown, Walter Tuttlebee, Peter Grant and many more. At Ofcom Peter Ingram, Mike Goddard, those in my R&D team and others have provided invaluable guidance. Finally, as always, thanks to Alison, my wife, who supports all my endeavours to write books with good humour and understanding.

Disclaimer

Note that the views and opinions presented in this book are those of the authors and not necessarily of the organisations which employ them. These views should in no way be assumed to imply any particular strategic direction or policy recommendation within the organisations thus represented.

I Emerging problems with the current spectrum management approach

1 Current spectrum management methods and their shortcomings

1.1 Why spectrum needs to be managed

A large and growing part of the world's output relies upon use of spectrum.[1] Frequencies are used both commercially, notably for mobile communications and broadcasting, and by public sector bodies to support national defence, aviation, the emergency services and so on. As demand grows spectrum needs to be managed to avoid the interference between different users becoming excessive. If users transmit at the same time, on the same frequency and sufficiently close to each other they will typically cause interference that might render both of their systems unusable. In some cases, "sufficiently close" might be tens or hundreds of miles apart. Even if users transmit on neighbouring frequencies, they can still interfere since with practical transmitters signals transmitted on one channel "leak" into adjacent channels, and with practical receivers signals in adjacent channels cannot be completely removed from the wanted signal. The key purpose of spectrum management is to maximise the value that society gains from the radio spectrum by allowing as many efficient users as possible while ensuring that the interference between different users remains manageable.

To fulfil this role, the spectrum manager provides each user with the right to transmit on a particular frequency over a particular area, typically in the form of a licence. Clearly, the spectrum manager must

[1] Spectrum is a term to describe a band of electro-magnetic frequencies. It is often used to refer to the radio spectrum, which extends from approximately 10 kHz to 300 GHz. This band is then subdivided with different parts being used for different applications. A licence will typically give a user the right to access or transmit on part of this spectrum, e.g. 800 MHz–820 MHz.

ensure that the licences that they distribute do not lead to excessive interference. In practice, this can be a highly challenging task.

This book is about how best the spectrum manager can accomplish this task, and in particular how the use of market mechanisms can assist them.

1.2 The current management mechanisms

Historically, the approach adopted by spectrum managers around the world to managing the radio spectrum has been highly prescriptive. Regulators often decide on both the use of a particular band and in some cases which users are allowed to transmit in the band.[2] Keeping a tight regulatory control over the use of the spectrum makes it easier for the regulator to ensure that excessive interference does not occur because the regulator is able to carefully model the interaction between neighbouring services and tailor the licence conditions appropriately. It also allows for other regulatory goals to be achieved – for example, ensuring that a service is available on a pan-European basis, or imposing coverage requirements to achieve ubiquity of services. Finally, it can result in high technical efficiency of spectrum use – that is to say in packing a large number of users into the spectrum. This is because like services in neighbouring bands tend to interfere less than unlike services and so can be allocated more closely together. If the regulator collects together like services and places them adjacent to each other it can increase the capacity of the spectrum (although maximising the capacity, or technical efficiency, is not always the same as maximising the benefits that society can gain from the spectrum, or economic efficiency, since the spectrum can be completely used but by a low value application).

As well as licensing users, the spectrum manager typically exempts other users from licensing. These exempted users are often assigned a band of spectrum sometimes known as unlicensed spectrum, or spectrum

[2] In the spectrum world, deciding the use of a band is called "allocation"; deciding which organisation can use it is called "assignment".

commons. The decision to exempt users is made on the basis that they will not interfere significantly with each other if they use the spectrum in an uncoordinated manner. In practice, this is likely only if they transmit at a relatively low power level such that the distance over which they can cause interference is small and hence the probability of there being another user within this small "coverage" area is low. Typical services that are exempt include cordless phones and wireless LANs such as WiFi. It is up to the regulator to decide which equipment to exempt, what the rules for its operation should be, how much spectrum should be set aside for its operation and where in the frequency band this should be.

The current spectrum allocation process operates at both a national and international level. International coordination is essential in some cases because the zones of possible interference extend beyond national geographical boundaries and in other cases because users are inherently international, e.g. aviation. Broadly, international bodies tend to set out high level guidance which national bodies adhere to in setting more detailed policy.

At the highest level of management sits the International Telecommunication Union (ITU), a specialised agency of the United Nations. The ITU's International Radio Regulations allocate the spectrum from 9 kHz to over 275 GHz to a range of different uses. In some cases these are quite prescriptive, e.g. "satellite". In other cases they allow substantial variation, e.g. "fixed or mobile". The Radio Regulations also set out how countries should coordinate with each other and in the case of global services, such as satellite, provides a mechanism for the assignment of rights to individual users. The ITU conducts the key parts of its business through World Radio Conferences which are typically held every three to four years. These are events attended by thousands of delegates from spectrum managers and users around the globe where potential changes to the Radio Regulations are considered. In some cases the ITU may seek international spectrum allocations for particular uses, for example in previous years it has allocated spectrum to global low Earth orbit satellite systems (of which Iridium is an example) and in its 2007 conference is intending to discuss whether there should be a global allocation for 4G cellular systems. Nothing in the Radio Regulations can constrain each country's freedom to manage spectrum as it wishes, as long as the impact on other countries is minimal and it is willing to accept the risk of interference.

In some countries, there are multi-national bodies coordinating across a region. For example, this is very much the case within Europe where the European Union (EU) and the Confederation of European Post and Telecommunication Agencies (CEPT) provide further harmonisation. Broadly, these bodies can be seen as local versions of the ITU, providing further coordination. Often their coordination is more specific, for example rather than simply designating a band as "mobile", they might designate it to a specific standard such as "GSM". Different bodies have differing levels of power. For the CEPT their decisions, like those of the ITU, are non-binding but if a country deviates from them it is expected not to cause interference to its neighbours as a result. However, the EU has legal powers and is able to require national spectrum managers under its jurisdiction to enact decisions. For example, EU Law requires national regulators in all member states to set aside spectrum in particular bands for GSM, although there is currently much discussion as to whether this decision should be repealed.

1.3 Shortcomings of the current system

The current approach "works" in so much as it licenses spectrum to particular users and ensures that excessive interference is avoided. This allows a range of uses of the spectrum in a stable and predictable environment. However, it is unlikely that it achieves the full objective of a spectrum manager of maximising the economic value derived from the spectrum. To do this, the regulator would need to make sure that spectrum was appropriately divided up between all the different possible uses and users in a way which maximised benefits to end users of spectrum using services.[3] Since it is almost impossible to predict the

[3] These services can be either commercial ones, purchased by firms and households, or public services such as national defence which governments "provide" on behalf of their citizens.

value that each different service provides under any given spectrum allocation it is difficult to see how a "command-and-control" approach to managing the radio spectrum could maximise value. However, it is possible that an extremely astute regulator might distribute spectrum in a manner that approaches this objective, at least in some bands.

In times where supply of spectrum exceeded demand, or where there were a relatively small number of services, it was more plausible that the regulator might achieve this goal. However, increasingly, demand for the spectrum has grown as has the number of spectrum-using services. There have been many pieces of evidence that suggest that regulators are failing to maximise value under such circumstances. Some examples are as follows.

- Some regulatory decisions, such as the allocation of spectrum to the ERMES paging system or the TFTS in-flight phone system in Europe, have resulted in spectrum being unused for over a decade. Clearly, it could have been put to an alternative use which would have resulted in some value.
- Widely differing valuations for the spectrum at auction, for example, the 3G auctions as opposed to spectrum auctions at 3.4 GHz suggest that the balance between different uses is incorrect. In this example, the much higher valuation of 3G suggests that there should be more spectrum made available for cellular, with perhaps less for fixed wireless or other applications.
- Many new applications or technologies have had great difficulty in gaining access to spectrum – for example the iBurst cellular technology or more recently Mobile TV systems. While it is not certain that these would increase the value of the spectrum, their difficultly in entering the market may be a symptom of an excessively rigid system.
- Some applications which have been granted spectrum free, such as aviation radar, have not modernised their radar systems for many decades despite the availability of much more efficient technologies, suggesting there are insufficient incentives for some users to optimise their use of the spectrum.

The current system is also becoming increasingly difficult for the spectrum manager to operate. They may receive frequent requests for new spectrum or to allow existing users to change application. They may also suffer complaints of unfairness as convergence of communications services allows some users who have accessed the spectrum for free to compete with other users who have paid for it.

All of this suggests that the current approach to spectrum management, whereby the regulator selects the use of the spectrum and in some cases the user, is probably not maximising the value of the spectrum and is becoming increasingly difficult for the regulator to administer.

1.4 Alternative management approaches

Economists have long argued that market mechanisms should be applied to radio spectrum. Seminal papers in this area start with Coase in 1959 [1]. The essential idea here is to allow pricing mechanisms to act as an incentive for holders of spectrum to optimise their use – buying more if their business case can justify it, selling spectrum if they have excess, and adopting new technologies that can use spectrum more efficiently where economically viable. Economic theory suggests that in a market which is performing well, this will lead to a division of spectrum that maximises economic value. Under such an approach the regulator sets out rules that enable markets to function while ensuring that interference is controlled and then takes a back seat, leaving it to the market to determine the use and users of the spectrum. However, the development of such rules is complex, as we will be discussing in this book.

The simplest of the market instruments to adopt is probably the use of auctions as a mechanism to distribute spectrum. Auctions are now used as the preferred mechanism for assigning spectrum in many countries and they solve the most pressing problems for many of the regulators by allowing spectrum to be assigned where demand significantly exceeds supply in a transparent and fair manner. But auctions on their own still "freeze" the assignment of spectrum. They need to be accompanied by mechanisms to trade and change the use of spectrum as market conditions change and new services become available.

1.5 How this book addresses the new approaches

This book analyses the key new approaches proposed for managing spectrum. It is structured as follows.

- Chapters 2 and 3 consider new spectrum-using technologies and the implications that they might have for the manner in which spectrum is managed.
- Chapters 4 through to 10 look at the application of market mechanisms to radio spectrum. Chapter 4 provides an overview as to how markets can be used, and then subsequent chapters consider particular economic tools; namely auctions and property rights. This part of the book concludes with a look at the possibilities for competition concerns to emerge and an assessment as to whether market intermediaries in the form of "band managers" might be beneficial.
- Chapters 11 and 12 consider how economic methods can be applied in those areas where market forces, for whatever reason, may not be appropriate. The main tool for achieving this is administratively calculated incentive based spectrum pricing.
- Chapters 13 and 14 look at the commons as an alternative approach to managing spectrum, considering how commons work and setting out guidelines as to where they might be used.
- Chapters 15 and 16 discuss how the analysis set out above needs to be adapted for public sector users and developing countries, respectively.
- Chapter 17 provides brief conclusions.

Reference

[1] R. H. Coase, "The Federal Communications Commission", *Journal of Law and Economics*, **2**, 1–40, 1959.

2 How changing technology is impacting spectrum management

2.1 Technology used to lend itself to discrete allocations

Until recently all technologies used a relatively narrow bandwidth and assumed that they were the sole users of that frequency. For example, the GSM mobile phone system transmitted signals with a 200 kHz bandwidth, which at 900 MHz is less than a thousandth of the carrier frequency. The systems were designed assuming that there would be little interference, and where there was it would be carefully controlled by the operator.

The result of the use of these technologies has been to regulate the spectrum by frequency. That is, the spectrum is divided up into discrete parcels of frequency, for example 915–925 MHz, and assigned to a particular user. That user then expects that they will be given exclusive use of the band.[1] This is shown diagrammatically in Figure 2.1, and has been the system on which spectrum management has been based for almost 100 years.

This approach facilitates the same use of spectrum in multiple countries, often known as harmonisation. By aligning the decision as to what the spectrum is to be used for across multiple countries, the same technology, such as GSM, can be deployed. This brings a range of benefits including economies of scale, international roaming and reduced interference. However, it also brings some disadvantages including the need for the regulator to predict the optimum service and technology and tying many countries to the same frequency plan regardless of whether the need differs from country to country.

[1] In practice, there are other emissions in the band, often from devices such as hairdryers or computers. These are allowed so long as they fall below recognised levels. Hence, the term "exclusive use" is not precise. We will return to this issue in much more detail at a later stage when we discuss ultra-wideband.

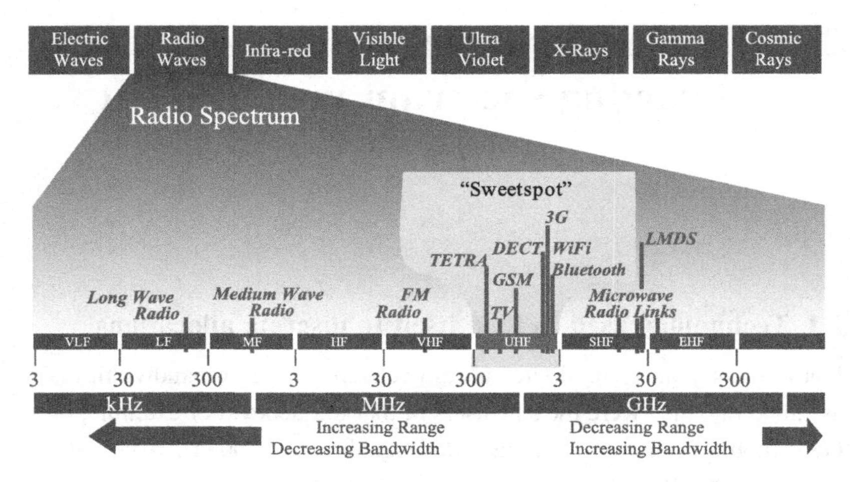

Figure 2.1. Schematic representation of the division of radio spectrum.

After over 100 years of radio spectrum management, the technology underlying the traditional approach to spectrum management is slowly changing. Two new technologies offer the potential for spectrum access under different conditions from today – these are cognitive radio and ultra-wideband. Other similar technologies might be expected to emerge. In addition, the ability to build multi-modal radios reduces the need for international harmonisation, changing some of the drivers for conventional spectrum management. These are discussed in more detail in the following sections

2.2 Multi-modal radios

A multi-modal radio is one capable of working across multiple bands and multiple technologies. To some degree multi-modal devices have existed for many years. A classic example is the AM/FM radio receiver, capable of working across both the medium wave (MW) and very high frequency (VHF) bands and capable of decoding signals with amplitude modulation (AM) and frequency modulation (FM). The increasing trend to multi-modality has been particularly noticeable with mobile phones. Early GSM phones worked only in the 900 MHz band. By the mid 1990s dual-band phones working at 900 MHz and 1800 MHz were available and by the late 1990s the "tri-band" phones were capable of working at 1900 MHz as well. By 2001 phones had added BlueTooth capabilities, using a completely different radio technology in the 2.4 GHz band. Then by 2003 phones were available that could also work on 3G systems, another completely different radio technology in the 2.1 GHz band. Over time, these extra capabilities have been added for relatively small cost increments and without noticeably changing the size, weight or battery life of the device.

The multi-mode device does not challenge the division of spectrum in discrete frequency elements. However, it negates to some extent the need to harmonise spectrum and technology standards between different countries. Now if different countries make different choices, as for example has been the case with the USA and Europe for cellular technologies, a single phone can still be produced which allows roaming across these countries and has sensible economies of scale. Of course, there is an additional device cost for multi-mode capabilities but in many cases this is sufficiently small that it becomes insignificant to the average consumer.

2.3 Cognitive and software defined radios

2.3.1 Introduction

Of greater significance is the potential for radios to move across the frequency band, seeking free spectrum and adapting themselves to use this spectrum. Such devices are sometimes called software defined radios (SDRs) and sometimes cognitive radios. But of the two "cognitive radio" is more accurate and is also the terminology gaining the most widespread use. Firstly, we address the terminology.

- **Cognitive radios**. These are devices which are "cognitive" of their surroundings. They can monitor transmissions across a wide bandwidth and note areas of spectrum that appear to be currently unused. They are intelligent enough to adapt their transmissions to the characteristics of the spectrum that they find vacant, using appropriate modulation and coding methods. They can detect when the spectrum is being used by its owner and move to a different band. Perhaps crucially, they can determine the most appropriate access to the spectrum without any central control from a base station.

- **Software defined radios**. Strictly an SDR is a device whose entire operation is defined by software. In purist terms this implies that the broadband signal captured by the antenna is digitised by an analogue-to-digital converter and from that point on all signal processing is performed using software. This would be radically different from current radios where filters and mixers are used to select and down-convert small parts of the spectrum prior to its analogue-to-digital conversion. Were such an SDR developed, it would have the advantage of being extremely flexible. A software library could contain information about multiple different technologies and new software could be downloaded to the device as technologies changed over time. Having an SDR would make the realisation of a cognitive radio much simpler, which is why the terms, rather confusingly, have been used interchangeably.

2.3.2 The progress of technology

Having now defined the terms it is worth addressing the progress of the technology. Just like the AM/FM radio of the previous section, cognitive radios have existed in a simple form for many years. The DECT technology, which was developed around 1990 and is in use in most of the cordless home telephones world-wide, requires the base station to monitor the ten frequency channels allocated to DECT and select the one with the lowest interference. The handset then moves to this same frequency. There are many other different radio technologies with similar characteristics. But these all fall short of becoming cognitive radios of the form currently envisaged because they operate only within their own pre-defined frequency bands.

True cognitive radios are technically possible, but expensive. In order to be able to work across multiple bands a cognitive radio needs a "front end" capable of scanning a wide range of frequencies. This implies that the antenna, filters, mixers and down-converters can work across multiple frequency bands. While technically possible, this adds cost and complexity to the radio. Antennas typically need to be larger as they cannot be optimised for particular frequencies but can then end up with unwanted directionality at certain frequencies. Often multiple filters must be employed, with switching between them, depending on the frequency selected. A good example of this can be seen by comparing simple FM/AM radios with those also capable of short wave (SW) reception. The former are sold for less than $10 each. The latter retail for around $100. The additional cost of a broadband front end depends on many factors, such as the actual range of frequencies over which it needs to be flexible, and will tend to fall over time as devices improve, but at the moment would likely render such devices unviable compared to existing cellular phones.

This flexibility could, in principle, be provided by SDRs, so it is worth briefly exploring their progress. We are still some considerable distance from being able to implement the classic SDR where the antenna output is digitised. Analogue-to-digital converters are still nowhere near fast enough to work at frequencies such as 900 MHz where much of the cellular traffic is concentrated. Digital signal processors (DSPs) and microprocessors are not fast enough to perform the necessary signal processing in real time. While processors and converters are steadily improving, the complexity of new standards such as 3G is growing. Indeed, it has been noted that the increase in complexity of 3G over 2G is actually greater than the growth in device capabilities predicted by Moore's law over the same period.[2] Were this to continue then the prospects for true SDRs would decrease rather than increase over time. This has led experts to predict that commercial SDR is unlikely to arrive before 2015 and may not materialise for many years after that.

As with all these new technologies, there are simpler versions that can be implemented. Even if the front end is not digital, it is possible to

[2] According to Moore's law the number of transistors on a chip doubles around every 18–24 months.

modify some of the software used for signal processing and protocol management. At its simplest, a handset that can download a new ringtone could be argued to be an SDR. SIM updates over the air are now widely used to change service parameters by software. Handsets that can have their protocol-layer software modified over the air are not too far away. This would provide the software adaptability needed for cognitive radio if not the broadband front end.

So, in summary, the technology for cognitive radios is still evolving and is not yet commercially realisable in a meaningful manner. However, the time when basic cognitive radios which cover only a limited range of frequencies can be implemented is not far away, and with some spectrum management decisions taking up to 10 years to implement, it is appropriate to consider the impact that cognitive radios would have once they appeared.

2.3.3 The implications of cognitive radios

Assuming for the moment that the technology behind cognitive radios could be developed and implemented commercially, and that the cognitive approach could be made to work, the implications for spectrum management might be highly significant.

Measurements seem to suggest that much of the radio spectrum is unused for much of the time, despite the fact that it has almost entirely been assigned. This is unsurprising in some cases. For example, emergency service organisations will often aim to leave a buffer of unused spectrum that can be called upon in an emergency situation. Alternatively, spectrum may be used in one part of a town, but not another. The proponents of cognitive radio suggest that so little of the spectrum is actually being used that, were cognitive radios deployed, there would no longer be a spectrum shortage. If there is no shortage then there is little need to regulate. Hence, proponents suggest that cognitive radios could lead to a world where all spectrum is unlicensed (the "commons" approach, see Chapter 13) and devices access spectrum on a dynamic and intelligent basis.

2.3.4 The problems with cognitive radios

There are problems with the scenario laid out above.

- In practice, spectrum may be more heavily used than it appears.
- Cognitive radios suffer from a fundamental "hidden terminal" problem.
- Some users will value certainty of access.

Each of these points is now considered separately, with the implications and likely outcome discussed in the next section.

Spectrum may be more heavily used than it appears. Simple monitoring of spectrum to determine occupancy suffers from many problems. Some examples are as follows.

- The transmitter may be hidden behind a hill or other obstacle with the result that the monitoring equipment cannot ascertain that the channel is in use.
- A technology with very low signal strength, such as GPS satellite transmission, may be being used, which falls below the noise threshold of the measuring equipment but nevertheless is detectable to GPS receivers.
- Frequencies are typically re-used between cells. This requires a gap between the cells using the same frequency to avoid interference. In this gap, the spectrum appears to be unused. However, were a transmission made it would result in interference to the existing use of the spectrum.
- Spectrum is often underused outside cities; however, this is because of a lack of demand and does not create a need for a different technology.
- Newer technologies employ adaptive antennas which steer the radio signal towards a receiver. Unless the monitoring equipment is in the same direction as the receiver it may not pick up the signal.
- Theory relating to the blocking of capacity in communications systems shows that channels typically need to have an average utilisation of less than 80% in order to provide capacity for peaks in traffic flows.

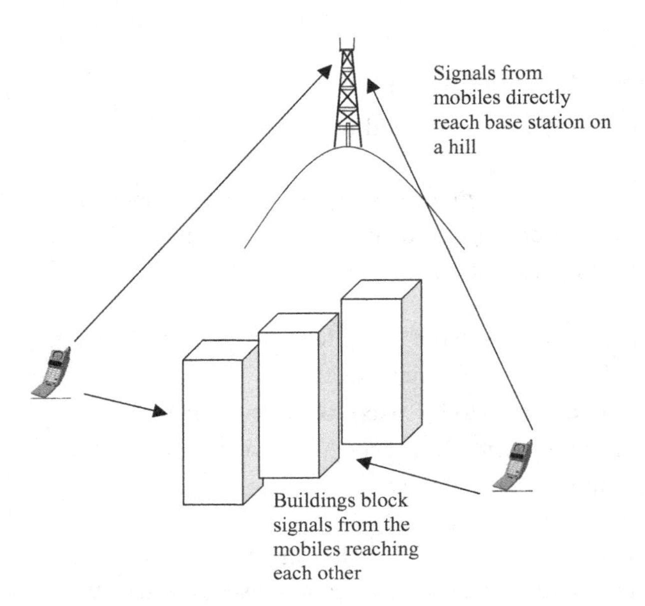

Figure 2.2. The "hidden terminal" problem.

These factors imply that the utilisation of the spectrum will always be greater than is detected through monitoring. It is far from clear what the "real" utilisation of the spectrum is, and some organisations are now embarking on larger scale programmes to try to determine this.

Cognitive radios suffer from the hidden terminal problem. The "hidden terminal" problem is shown in Figure 2.2. A cognitive radio might make a measurement and not spot any activity on a piece of spectrum. However, there might be a legitimate user of that spectrum behind the next building, transmitting to a tower on the hill. Because the building is between the users, the cognitive radio does not receive the legitimate signal and so concludes the spectrum is unoccupied. But because both users are visible to the tower on the hill, when the cognitive radio transmits its signal is received as interference at the tower.

This problem is solved by the tower on the hill transmitting a signal indicating whether the spectrum is used. A terminal then requests usage of the spectrum and, if granted, the tower indicates that the spectrum is busy. Such an approach works well but it requires central management by the owner of the band which has significant implications for spectrum management as discussed below.

Some users will value certainty of access. The cognitive radio approach relies on finding unused radio spectrum. In some cases, it might be difficult to find this spectrum with the result that the cognitive user may be unable to transmit. For some users this may be unacceptable. Hence, there might remain some bands with greater certainty of access where normal access approaches were used. If the probability of cognitive access fell too low it would become of little value to most users.

2.3.5 The likely outcome for spectrum management

As has been demonstrated above, there are still many uncertainties associated with cognitive radios, including the technology, whether they will work in practice and the amount of spectrum that is available to them. As some of these get resolved, the most appropriate spectrum management response towards cognitive radios might evolve.

At this stage, the hidden terminal problem suggests that cognitive radios might best work where the owner of the spectrum broadcasts some form of beacon signal to indicate whether cognitive radios can access the spectrum. They then might charge some fee for the right to make a cognitive access to their spectrum. The role for a spectrum manager in such an environment is limited. By providing licence holders with the right to sub-lease their spectrum this form of access would become possible. Sub-leasing is considered by many to be part of the package of rights needed for spectrum trading.

The regulator could go further, for example, as follows.

- They could help promote standards for cognitive access, perhaps standardising the form of beacon signal provided.
- They could inform the owners of radio spectrum about cognitive access and provide them with the information they need to add this capability to their system.
- They could mandate that owners of particular bands provide cognitive access.

Whether regulators would want to do any or all of these is not clear. If regulators tend towards market forces to manage radio spectrum then mandating cognitive access seems unlikely.

2.4 Ultra-wideband

2.4.1 Introduction

In transmitting a radio signal it is always possible to trade bandwidth against signal strength. That is, if more bandwidth is allocated for a signal it can be transmitted over the same range with a lower signal-to-interference ratio. The bandwidth of the signal is increased by "spreading" it. This tends to involve turning each bit of information to be transmitted into a pre-defined stream of bits, resulting in more data and hence a higher data rate and bandwidth. The receiver can then look for this known stream of bits which is easier to find than a single bit. This is the principle on which CDMA transmissions are based.

Ultra-wideband (UWB) takes this approach to its extreme. Signals can be spread across much of the bandwidth between around 3 GHz and 10 GHz (although many variants on this band exist). With such a large increase in bandwidth, the power levels can be correspondingly reduced to such a low level that they fall to the limit placed on unwanted emissions from non-communications devices.

Because of the way that UWB is implemented, it can be used for a range of purposes. As well as communications it can provide accurate "through the wall" radar detection or ground probing radar and highly accurate short range positioning systems.

2.4.2 Technological issues

Implementation. Unlike cognitive radio, UWB is very simple to implement. The classic UWB implementation generates a stream of short pulses and then modifies the timing of each of the pulses to convey the information. Generating pulses is relatively simple and when coupled to the low power levels associated with UWB this results in very low cost devices that can easily be incorporated into larger chipsets or added to other electronic systems such as computers or PDAs. The technology is already available to do this. Hence, there are no significant technological obstacles to overcome.

Range. Range is a problem for UWB. In order to keep power levels below the threshold for unwanted emissions, even with the bandwidth gain, the range of UWB falls to around 10 m. Hence, the main applications for UWB are for personal area networks, or wireless networks within a room. One application often mentioned is the wireless transmission of signals from DVD players to screens and speakers within the same room.

Interference with other systems. One of the key issues is whether UWB transmitters will interfere with existing systems. If they do not, then there would seem to be little issue with allowing UWB. If they do, then a decision on UWB depends on issues such as the licence conditions for the current licence holders and whether the net benefit of UWB is likely to exceed the overall cost of interference. As is often the case, there is no simple answer to the question of interference. The likelihood depends on the sensitivity of the existing technology to interference and to assumptions as to the likely number of UWB transmitters in the vicinity. For example, on the one hand, UWB transmissions would be highly unlikely to interfere with broadcast TV reception through a rooftop antenna as very little UWB signal would likely reach the antenna. On the other hand, UWB might interfere with a 3G mobile which itself operates with broadband low-power signals and where multiple UWB devices might be in the same room as a 3G handset. Even where there is interference, the impact might not be to prevent communications but only to reduce its effectiveness. For example, in the case of 3G, the UWB interference might manifest itself as a reduction in overall system capacity rather than a blocking of specific devices. Overall, studies show that under some conditions, UWB might provide some level of interference to some users of the radio spectrum. Making a decision on UWB when faced with such an unclear answer to the question of interference is problematic.

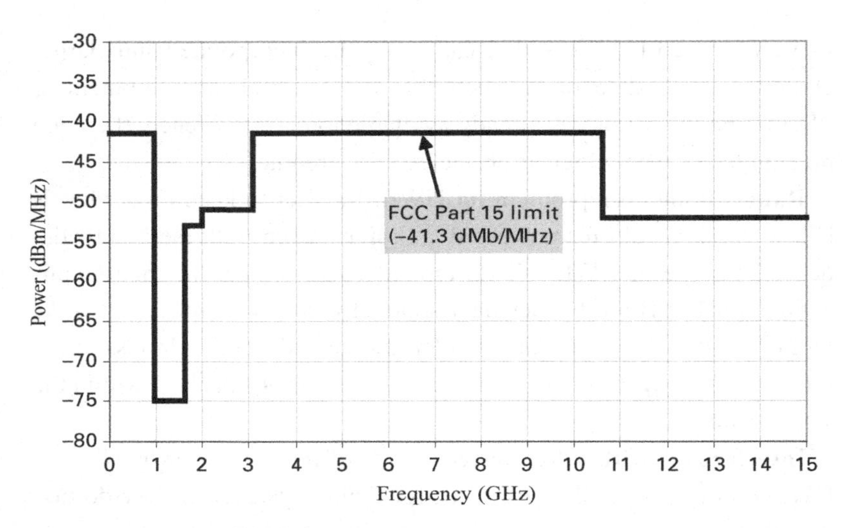

Figure 2.3. Frequency mask for UWB.

2.4.3 Current international position

At the time of writing in 2006 only one regulator had decided its position on UWB. The FCC in the USA had decided to allow UWB but with the restriction of a particular frequency mask, shown in Figure 2.3.

Broadly, the FCC have permitted UWB in the frequency band 3–10 GHz up to the same power levels as unwanted emissions from non-communications devices (the "FCC Part 15 limit"). They have particularly protected GPS transmissions at around 1 GHz taking into account the fact that this use is especially sensitive to interference.

Generally setting maximum power levels at the Part 15 limit has much to recommend it. If non-communications devices are allowed to emit at these levels, then there is a reasonable argument that communications devices should be allowed to do so as well. The "notch" for GPS is also practical in that it protects an existing service. However, it indicates that some services require more protection than others and implicitly prevents services like GPS being deployed in frequency bands between 3 GHz and 10 GHz.

Other regulators are currently considering their position on UWB, with the EC looking likely to issue a directive harmonising use across Europe. The regulations imposed in Europe look likely to be significantly more restrictive in the bands below 6 GHz than those in the USA, which may persuade manufacturers to concentrate on devices above 6 GHz.

2.4.4 The likely outcome for spectrum management

On the one hand, regulators could consider UWB a one-off technology, allow its use, and continue in all other respects to manage spectrum in the manner that they have done in the past. This is the position that the USA is adopting at the moment.

On the other hand, the licensing of systems based on power levels rather than frequency bands could be seen as such a radical change to spectrum management that it necessitates a complete re-think in the way that spectrum is managed.

These are issues of such complexity and importance that we will return to them in the next chapter.

2.5 Summary

In this chapter we have considered the impact of new technologies on spectrum management and discussed the following.

- Multi-modal radios are gradually reducing the advantages of international harmonisation, making it easier for regulators to allow the use of market forces.
- Cognitive radios potentially offer the promise of a radical change in the way in which spectrum is used, but in practice may be best enabled simply by providing spectrum owners with sub-leasing capabilities.
- UWB is a radically different technology in the manner in which it uses spectrum and may require a significant re-think of spectrum management principles.

3 Alternative ways of dividing spectrum

3.1 Spectrum has been divided by frequency

In this chapter we look in more detail at different mechanisms for dividing up access to the spectrum. The previous chapter noted that spectrum is typically divided by frequency, with each user being given exclusive access to a frequency. In this chapter we look at the current process in more detail, assess all the different mechanisms for dividing spectrum and look at the impact that a change would have on the current use of radio spectrum.

Before delving deeper into spectrum access it is worth considering the question of what spectrum actually is. When a user gains a licence they do not actually gain any "spectrum". Indeed, there is not really any such thing as radio spectrum, it is merely a representation of electro-magnetic radiation. Instead, users typically get a licence to transmit at a specific frequency and often with other conditions attached. Implicit to this is the expectation that they will be able to receive the transmitted signal without any harmful interference. We return to look at these conditions or "rights" in more detail in Chapters 7 and 8 where they become an important component of spectrum trading. As long as these rights are not compromised, other users could in principle access the same frequencies. Any method of dividing access to the spectrum that allows users to receive their transmitted signal would be viable.

Within traditional management structures the key types of spectrum access are:

- exclusive access,
- geographical sharing,
- band sharing.

Exclusive access

The simplest of all forms. A user is given a licence to transmit on a particular frequency throughout the whole jurisdiction of the regulator (normally a country). No other user is given access to the same spectrum with the exception of unwanted emissions while their impact remains small.

Geographical sharing

In this case a user is given an exclusive licence but only for part of the country. A different user is given a licence to use the same frequencies in a different part of the country. The regulator needs to be sure that the two users are sufficiently far apart that they will not interfere with each other on the boundary of their operations. In practice, this implies either a guard zone between coverage areas or some restrictions on emissions near the boundary.

Band sharing

In this case, different uses are allowed in the same spectrum. A typical example is satellite transmission and fixed links. Because both users have directional antennas and as the satellite antennas point somewhat vertically whereas the fixed link antennas point horizontally, there is a fair degree of isolation between them. However, it is still necessary to maintain some geographical separation. As a result, band sharing schemes are often essentially methods of geographical sharing where the difference in application allows the geographical separation to be smaller than might otherwise be the case.

3.2 UWB raises the possibility of division by power

As discussed in Chapter 2, UWB technology works somewhat differently from other technologies in that it can use the same spectrum but at a lower power level. On a spectrum assignment map this can be represented as shown in Figure 3.1.

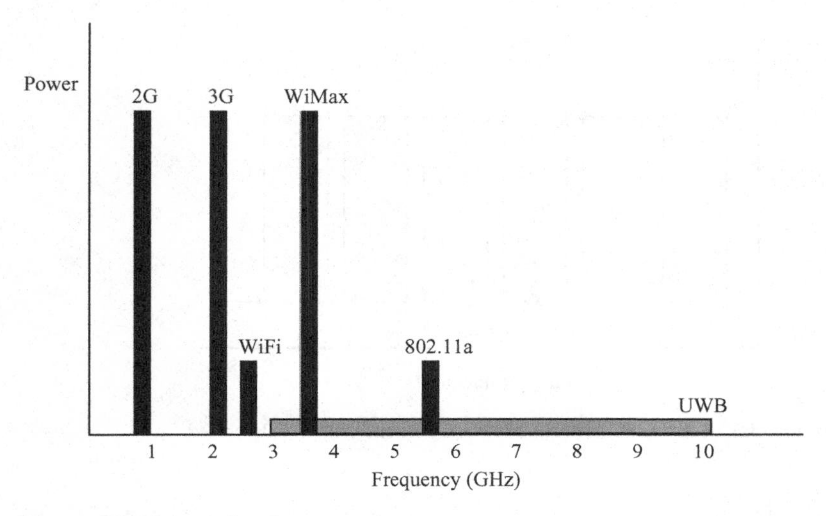

Figure 3.1. Schematic of spectrum/power.

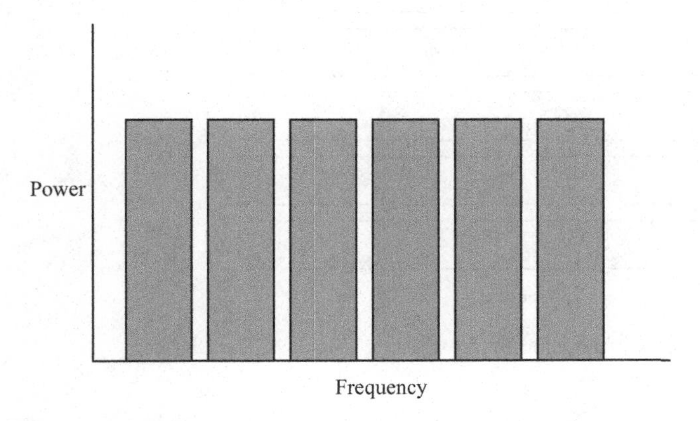

Figure 3.2. Simplified diagram of frequency assignment.

If we simplify this figure then the current assignment by frequency could be shown schematically as in Figure 3.2.

If UWB were allowed, this would result in the situation shown in Figure 3.3.

This representation has led many to suggest that an alternative to division by frequency might be a division by power. This could be represented as Figure 3.4.

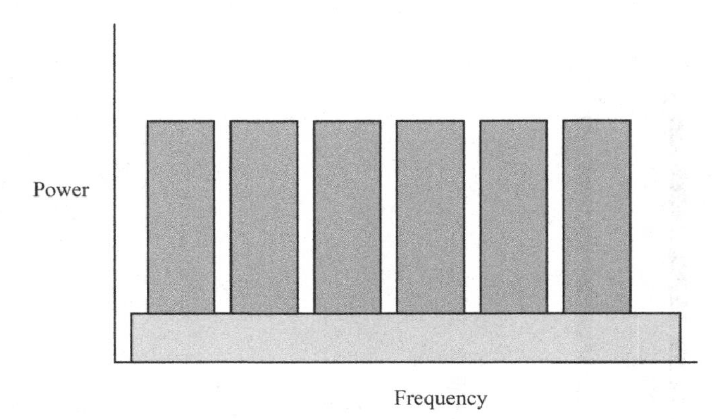

Figure 3.3. Schematic of frequency assignment with UWB added.

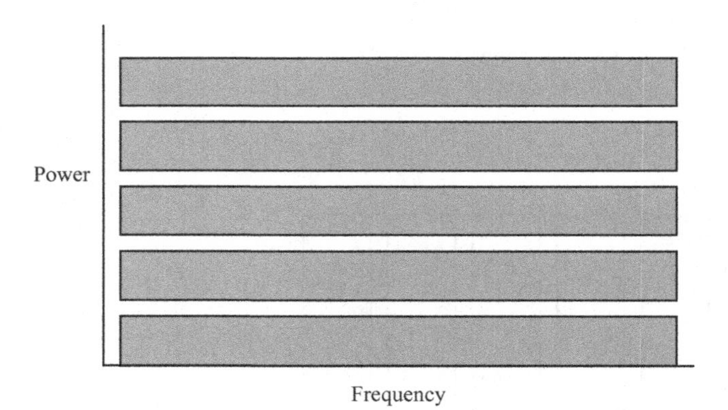

Figure 3.4. Schematic of division by power.

Such a division is not possible. Power is not the same sort of variable as frequency. It is possible for one user to transmit on one frequency and not to affect a different user, using another frequency.[1] That is, transmission on one frequency need not conflict with transmission on a separate frequency. The same is not true of power. All users transmitting on the same frequency will interfere with each other regardless of their power level. Allocating one user a maximum power level of,

[1] This property is known as "orthogonality".

for example, 30 dBm and another 40 dBm will simply result in them interfering with each other. Assuming all the users operate in the same geographical area then the effect will be that the user with the lowest power level will be operating with the most negative signal-to-interference ratio. To be able to receive information under this condition they will need a large spreading gain such that they can make use of the inherent redundancy this provides in order to search for their signal amongst other stronger signals. This implies that they will only be able to transmit at a relatively low data rate as the combination of low rate and high spreading factor will fill the overall frequency allocation that they have been given. Higher power users will be able to have higher data rates and lower spreading factors because their signal-to-interference ratios will be higher.

The net result would be mostly the same as if each had been given exclusive access to particular frequencies, with some users getting more than others. Perhaps this should not be a surprise – the physical laws that underlie the amount of information that can be sent are unchanged by the mechanisms used to divide spectrum. The only difference is that in the case of division by power, if one user stops transmitting temporarily then the others can increase their data rate to take advantage of the reduced interference. The equivalent of this in the case of division by frequency would be the cognitive radio. However, it is possible that under typical usage conditions, division by power might be more efficient than division by frequency. This is something we return to in more detail below.

A mix between the two approaches is also possible, for example, a power–frequency allocation like that in Figure 3.5 could be adopted. This would allow particular users to have the interference mix that suited them well, but would be extremely complex to administer and would not necessarily result in any significant improvement in spectrum efficiency.

So division by power would be more efficient than division by frequency only if it were better at allowing spectrum temporarily unused by one user to be used by another. Whether this is so is not a question that can be answered definitively because it depends on the

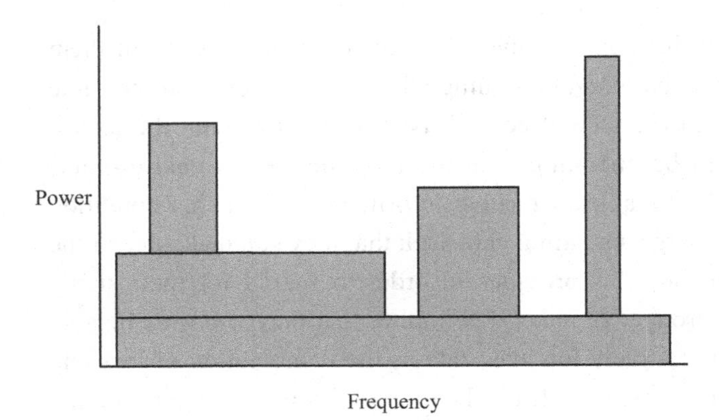

Figure 3.5. Schematic of a mixed frequency and power division approach.

extent to which more advanced technologies are deployed in the division by frequency approach. For example, if cognitive radios were deployed then they might be effective at making use of unwanted spectrum. Alternatively, operators in the same band could agree on an overall trunking mechanism to divide their resources such that they all had access to the whole spectrum in some controlled manner.

Division by power might have inefficiencies associated with it. Commercial systems that effectively use a form of division by power are termed code division multiple access (CDMA) systems. These include most of the proposed 3G technologies. These systems maximise capacity when the maximum degree of orthogonality is achieved between users. This is done by ensuring that the spreading codes used by each user are both different and have specific properties which maximise orthogonality. So firstly, in a division by power system, capacity would be maximised if all the different users coordinated their allocation of spreading codes – a level of coordination akin to the trunking approach mentioned above as a way to maximise frequency usage. However, despite aiming for orthogonality, even the best designed systems rarely achieve it perfectly in practice. This is because the propagation environment tends to disperse the signal, resulting in multiple copies of the same signal being received at different times at the receiver. This phenomenon is known as multipath propagation.

Orthogonality depends on time synchronisation of signals, but multipath reduces the degree of orthogonality because only one of the received copies of the signal can be perfectly time synchronised.

Work on CDMA systems showed that such systems are most efficient when the interference is evenly distributed among all users. If the interference distribution is uneven then the system will typically be designed so that the user with the highest interference is still able to communicate. The result of this is that the less interfered user has extra "margin" that is unnecessary and hence inefficient. This lowers the overall system capacity. Single CDMA systems are very effective at spreading interference evenly. Division by power would be less efficient since any particular user would need to have enough spectrum and power to be able to communicate even when all the other users were also communicating. However, all the other users communicating simultaneously might be a relatively rare event and so there would be some inefficient margins built into the system.

So, in summary, although a division by power is possible, it would probably not provide any significant advantages over the current approach of division by frequency. Given this conclusion is it still appropriate to consider a UWB underlay – i.e. the approach shown in Figure 3.3 above?

This is not a question that it is possible to resolve fully here, and we will return to it in later chapters as we look at the implications on property rights of UWB. Against UWB is the fact that it will raise the interference level for existing users, albeit marginally in most cases, potentially making their use of the spectrum less efficient. In favour of UWB is the fact that interference already exists from non-communications devices, so that the presence of UWB may not make any material difference. Were there no emissions from other non-communications devices then UWB might significantly raise the noise floor for other users, some of which have quite sensitive applications. We can then rephrase the question to ask that, given the communications systems are already designed to tolerate some unwanted interference, and hence typically do not make use of the spectrum resource below this power level, why not allow valuable use through UWB? Effectively, spectrum is already divided up in the manner shown in Figure 3.3. As a result of this, UWB simply enhances the efficiency of the current approach. That is still not to say that it should automatically be allowed as a result since spectrum owners will increasingly find ways to work down into the noise floor region using adaptive technology and UWB takes some of this capability away. But the question is more of an economic one than it is a technical one.

3.2.1 Interference temperature

In November 2003 the FCC issued a notice of proposed rule making on a concept they termed interference temperature. The concept encompassed a range of different ideas and was never stated particularly clearly. However, the basic idea was that each licence holder should be given a set noise floor. Noise is normally measured in terms of signal power, but it can be specified in terms of the natural noise that would result at a particular temperature since natural noise rises with temperature. So specifying a temperature is another way of setting a noise level in terms of power flux density.

Once the user had a noise floor, the FCC postulated that if the noise they were experiencing was actually lower than the noise floor then unlicensed devices might be allowed to transmit up to the point that the noise floor was reached. This approach is somewhat similar to UWB, but UWB is not intelligent in that it does not assess existing noise conditions.

However, there is a fundamental problem with the concept which is that it is very difficult to measure the noise level. This is because there is normally a licence holder transmitting as well. A radio receiver can only with great difficultly, and poor accuracy, separate the licence holder's signal from other signals and determine the level of noise or interference in place. The FCC proposed that networks of monitoring devices might be used to measure noise levels across a wide area and signal this information through a beacon network, but as well as the difficultly in establishing the noise level this also requires costly infrastructure to be deployed when the value of the service gained is relatively low.

The FCC has issued no further documents on interference temperature since 2003 – it appears that in the face of these difficulties it has decided not to pursue the concept.

3.3 Other divisions are also possible

Many other means of dividing spectrum access could also be envisaged including:

- time,
- angle,
- polarisation.

Time

Different users could be given access to the spectrum at different times. Indeed, this is broadly the approach that would be followed were cognitive radios to prove possible. Arranging a division by time would be relatively simple.

Angle

It is possible to distinguish between signals arriving from different directions by using directional antennas. So it is quite possible, for example, to license two users of fixed links to use the same frequency band at the same time, terminating at the same point, as long as both links arrive at that common point from sufficiently different directions that the antenna directivity can reject the unwanted signals. Discrimination through direction of arrival can be used in three dimensions, as was hinted at in the earlier example of satellite systems sharing with fixed links. Arranging this may require detailed modelling on a link-by-link basis coupled to an understanding of the technical parameters of the equipment being deployed.

Polarisation

Polarisation relates to the plane in which the electrical and magnetic waves travel. It depends on the orientation of the antenna – a vertical antenna will produce signals with a different polarisation from a horizontal one. Signals transmitted with horizontal or vertical polarisation are mutually orthogonal to each other. Hence, in principle, two users could share the same location and frequency as long as they employed different polarisations. The problem with polarisation is that it changes on reflection with objects that are not perpendicular to the wave-front. For uses such as mobile telephony, the polarisation of a signal is almost completely lost through the multiple reflections that signals typically undergo before arriving at the receiver. In the case of fixed links this is less of an issue as there is often a line-of-sight and no significant reflections. Hence, depending on the usage, it might be possible to divide access by polarisation. In practice, at present users are typically not divided by polarisation, but some users choose to make use of both polarisations themselves to increase the capacity of the spectrum they have been awarded.

3.4 Summary: in practice, changes to spectrum division would be minor

Most of these alternative methods of dividing access to spectrum can be accommodated as a subdivision within the current overall process of division by frequency. For example, after a division by frequency, it is possible to further subdivide by time, angle, polarisation, geography or use. The regulator could choose to make this division itself, or could provide the licence holder with the freedom to do so, perhaps through the flexible type of licence envisaged under trading.

Division by power could be different. As explained above there are complex means whereby users could be given different power levels to transmit on. But we also explained that this did not seem worthwhile and that UWB fitted well into the current system of spectrum division. Major changes to these approaches look unlikely.

II Markets

4 Market solutions

4.1 Introduction

Historically, regulators have assigned frequencies by issuing licences to specific users for specific purposes – an administrative approach. The administrative approach can also be more or less prescriptive on the details of spectrum use. Often it has involved specifying what equipment a licensee can use and where, and at what power levels it can be used.

This is a good way to control interference, yet such methods are often slow and unresponsive to new technological opportunities. They also assume a level of knowledge and foresight on the part of the spectrum regulator which it may not possess. Attention has recently been focussed on creating genuine markets for spectrum and spectrum licences under which both the ownership and use of spectrum can change in the course of a licence's operation. This is a major step beyond the auctioning of licences which are not subject to trading and change of use. It does, however, require the full specification of what "property rights" to spectrum can be traded and utilised.

Some spectrum, especially for short-range use (BlueTooth, RFIDs, microwave ovens, various remote control devices, wireless security systems, etc.) need not be licensed at all. This might be the case where users do not interfere with one another, or because new technologies can be employed which are capable of dealing with interference as it happens. If such coexistence can be achieved, the spectrum commons approach is desirable.

Regulators should look for the right balance among these three methods of administrative assignment, use of markets and commons. The choice will be based on such things as the general scarcity of spectrum in various parts of the country and in various portions of the spectrum; the human and financial resources available to the regulator;

Table 4.1. *Division of allocation as proposed by Ofcom*

	Percentage of spectrum allocated in	
Spectrum allocation method	2000	2010
Administrative	96%	22%
Market	0%	71%
Commons	4%	7%

Source: Ofcom (http://www.ofcom.org.uk/consult/condocs/sfr/sfr2/)

the characteristics of the technologies used, for example whether they are localised or wide-area; and opportunities for innovation and commerce. The growing recognition that spectrum regulators may not be able to collect and process the information needed to make efficient administrative assignments is one of the factors promoting spectrum reform throughout the world.

As an illustration of the changing balance among methods of spectrum management the United Kingdom spectrum regulator, Ofcom, is proposing a radical shift from administrative methods to a market-based approach, and a smaller expansion of the commons, over the period up to 2010, as shown in Table 4.1.

4.2 Market methods

Market methods are being employed both at the primary issue of spectrum licences, when auctions are used and, more significantly, by allowing spectrum rights to be bought and sold in the lifetime of a licence (trading) and allowing a change of use of the relevant spectrum (sometimes called liberalisation).

Trading involves the transfer of spectrum usage rights between parties – either the government or regulator and a public or private licensee (at primary issue), or between two licensees, through so-called secondary, or post-issue, trading. We explore the issue of defining the property rights to be traded in more detail in Chapters 7 and 8. Trading can lead to a more economically efficient use of frequencies. This is because a trade will take place only if the spectrum is worth more to the new user than it was to the old user, reflecting the greater economic benefit the new user expects to derive from its use. In the absence of misjudgements or irrational behaviour on the part of the buyer or seller, and if the trade does not cause external effects, then it can be assumed that spectrum trading contributes to greater economic efficiency.

As well as this direct effect, which at the same time boosts transparency by revealing the true opportunity cost of the spectrum, secondary trading also results in a series of indirect positive effects. Spectrum trading makes it possible for companies to expand more quickly than would otherwise be the case. It also makes it easier for prospective new market entrants to acquire spectrum in order to enter the market.

If the introduction of spectrum trading is combined with an extensive liberalisation of spectrum usage rights, there will be a considerable incentive for incumbents to invest in new technology in order to ward off the threat of new entrants in the absence of other barriers to entry (i.e. the unavailability of spectrum). This in turn will boost market competition. These economic efficiency gains will not be realised, however, if transaction costs are too high or if external effects intervene (particularly, anti-competitive behaviour and interference). Transaction or administrative costs can be kept low by ensuring, for example, that there are few bureaucratic obstacles to the transfer of spectrum. At the same time, there should be a source of clear information that allows prospective spectrum users to find out which frequencies are available, what they can be used for, who is currently using them and what needs to be done in order to obtain a right of use. To this end, it is advisable to set up a central database, which, for practical purposes, should be the direct responsibility of the spectrum regulator.

These criteria constitute the framework for a whole raft of institutional arrangements that determine the precise form of spectrum trading and set forth exactly how rights of use can be transferred. Institutional arrangements stipulate precisely who can make what decisions, when they can do so and under what conditions. They also set forth the implications this will have for the parties involved. Ideally, such a system will include full details pertaining to all aspects of spectrum transfers and trading. At the same time, one of the aims of any spectrum trading regime should be to keep transaction costs down. In practice, the vast quantity of important details means that both primary legislation and secondary legal texts are limited in terms of how far they can specify actual arrangements.

4.3 Market failures

Policy aimed at introducing market methods seeks to create markets in which prices are as close to costs as possible and where consumers can choose from a wide range of services. Fully effective competition is usually only possible where there are competing infrastructures, yet the spectrum-using technology, the markets which they serve and scarcity of radio spectrum create restrictions which often mean that an oligopoly, a market served by a small number of competitors, is the only possible stable outcome.

Highly concentrated markets for spectrum-using services often produce adverse consequences for end users. The worst form of such market failure associated with market power is a monopoly. Anti-competitive behaviour, in the form of an "excessive" acquisition of spectrum, can be prevented in different ways by the spectrum regulator, which can set spectrum caps or establish rules that specify how spectrum trading should take place, including prior approval of trades or transfers of spectrum. Equally, a country's competition regulator (if one exists) can attack abuses of power in markets for services which use spectrum, such as broadcasting services.

The above remarks clearly show that, even under a more flexible regulatory regime promoting use of market mechanisms, issues of market power will continue to be important. This, however, is not a reason to reject market methods. In fact, a more flexible approach to spectrum regulation, which not only allows multiple transfers of spectrum but, moreover, is also accompanied by a far-reaching liberalisation of usage rights, would actually tend to diminish rather than amplify potential problems of market power.

4.4 Conclusion

In summary, in order for a market to work it is necessary for the regulator to introduce the following.

- **Auctioning**. A mechanism for the regulator to ensure that any newly released spectrum into the market is acquired by those who value it the most (see Chapter 5).
- **Spectrum trading and liberalisation**. A mechanism for users to buy and sell spectrum amongst themselves, either to provide the same service or to provide any other service (see Chapters 6–8).

These tools are necessary and sufficient to enable market-based spectrum management. With these tools in place, the market is able to decide on the use and ownership of spectrum and there should be little need for regulatory intervention. The role of the regulator then becomes one of policing the spectrum, and acting directly in those cases where, for various reasons, market forces cannot be applied. These may include areas where international harmonisation is required, such as satellite use or low frequency communications which may travel thousands of miles.

The introduction of market mechanisms raises the possibility that in some cases there might be market failure – that is, the market mechanisms alone might not result in the regulatory goal of enhancing the efficiency of use of the spectrum. The greatest concern relates to possible anti-competitive actions which might include the hoarding of spectrum, the setting of excessive prices or degraded service quality. The implications of anti-competitive behaviour are assessed in Chapter 9.

Finally, in many markets such as the market for land or securities, intermediaries emerge – in the case of land in the form of real estate brokers and in securities in the form of traders. Chapter 10 assesses whether such intermediaries, known as band managers, might emerge in spectrum and if so what form they would be most likely to take.

5 Auctions

5.1 Introduction

Over the course of time radio spectrum administrators have applied many different ways to assign radio spectrum rights to users. Until the late 1980s administrators assigned licences using administrative processes that included lotteries, beauty contests[1] and awards on a first-come first-served basis. In the early 1990s a few administrators chose to auction spectrum rights, and following the large revenues raised in auctions for mobile telecommunications spectrum rights in the United States in the mid 1990s, interest in using auctions to assign frequency rights increased markedly around the world.

In the 1990s auctions appealed to some radio administrators, as it was felt an assignment process based upon market signals would reflect more accurately the value of spectrum and lead to more efficient use of spectrum, see FCC [1]. Advocates of auctions have long argued strongly that the outcomes in well-designed auctions are better for society than administrative procedures.[2] It is widely argued that the superiority of auctions stems from their objectivity and transparency.

Some radio spectrum administrators believe auctions ensure that the frequency rights go to those who should best own them because frequencies are typically granted to those willing to bid the highest amount. In competitive market economies, scarce resources are allocated efficiently if they flow to those willing to pay the highest amount

[1] Beauty contests are based on comparative selection and may involve hearings or the submission of detailed applications which are then scored according to rules devised by the radio administrator. The winner of a beauty contest is the applicant achieving the highest score.

[2] The Nobel Laureate economist Ronald H. Coase advocated over 40 years ago for the use of auctions in the United States as a superior means of assigning radio spectrum rights.

and radio spectrum is no different to any other scarce resource in this respect.

Nevertheless, the application of auctions as a mechanism to assign frequency rights has been challenged. Some commentators claim that auctions, by allegedly forcing bidders to pay high amounts, necessarily lead to higher final prices for services which rely on spectrum as an input. Such views are erroneous as discussed below, though that is not to say every auction is well designed and achieves the best for society. Indeed, some auctions have been poorly designed and badly implemented.

Except in a poorly designed auction, bidders are rarely forced to pay too much for radio frequency – after all, bidders in spectrum auctions should base their bids on the perceived value of the frequencies auctioned. However, in many cases the value of the frequencies on offer may not be known precisely, particularly in cases where the radio spectrum is intended to be used for new and untested commercial services, and so ex post a bidder may regret having paid too much for spectrum.

The value of spectrum has two components: a *private value* component and a *common value* component. It is the common value component and uncertainty surrounding this that can lead to bidders regretting certain actions in an auction for radio spectrum. The common value component is a value which is the same across all potential buyers of the spectrum. An often cited example of an object having a common value that is frequently auctioned is an oil tract lease. In such an auction, bidders in possession of the same information would likely hold the same value for the oil tract to be auctioned. The private value component is where prospective spectrum owners possess the same information but form different valuations reflecting "private" differences. For example, different firms may possess different patented techniques for using spectrum that leads to them holding different values for the same spectrum.

In auctions for spectrum which typically feature significant common value elements, bidders can suffer from the *winner's curse*: a situation where a bidder as a result of succeeding in an auction discovers that he or she has paid too much for the spectrum and has not taken into account information held by the other participants in the auction. It is claimed that the winner's curse does arise quite often in practice, though this can be mitigated by using certain auction designs, such as ascending bid auctions, in which there is a gradual price discovery process.

It is sometimes argued that auctions, and particularly those in which bids ascend over time, could encourage bidders to over-value spectrum, resulting in spectrum prices being too high. However, this scenario is exceptional and it is unclear whether bidders in spectrum auctions would over-value frequencies simply because of auction design. In cases where bidders have acknowledged afterwards that they over-paid for radio spectrum, as in a number of the European 3G auctions in 2000 and 2001, the over-valuation stemmed largely from optimistic expectations about future demand and supply conditions in the market at the time of the auction – rather than from the auction designs.

Another common fallacy levelled against auctions is the assertion that bid prices necessarily result in consumers paying more for services reliant upon radio spectrum. In well designed auctions bidders are required to pay for spectrum up front. This means that successful bidders will treat the cost of spectrum as sunk. It is well known from economic theory that sunk costs do not influence market prices. That is not to suggest auction prices are irrelevant, they do matter for those businesses buying spectrum and can make all the difference between a successful business model and a failure. But when it comes to the factors which determine the market prices paid by consumers for services that are supplied using radio spectrum, past auction fees do not feature. This point has been made on many occasions by economists. For example, Kwerel [2] has shown for the case of cellular spectrum auctioned in different regions in the USA that prices for cellular services did not vary as widely as prices for radio spectrum and that there was no statistically significant correlation between auction fees and the prices paid by consumers for mobile services.

The factors which influence bid values in an auction are illustrated in Figure 5.1. The key factors are the variables which determine the shape of the market structure in the post-auction phase. For example, the

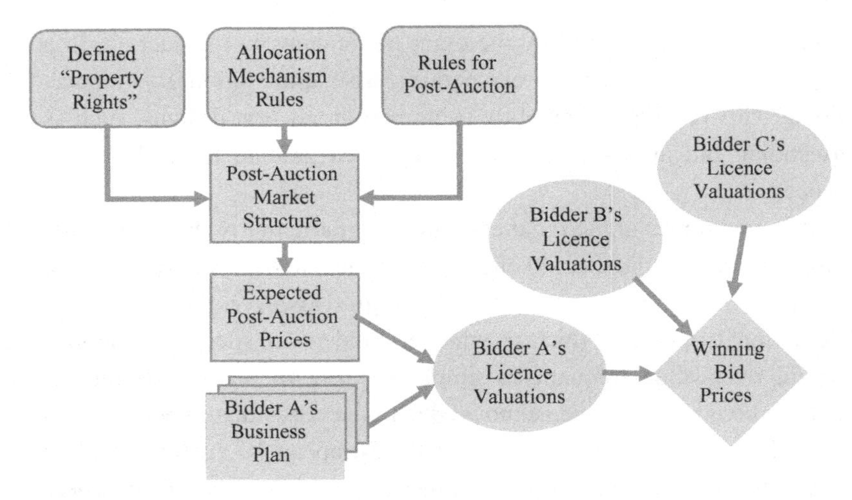

Figure 5.1. Factors influencing maximum bid values in a spectrum auction.

number of competitors in a market will be influenced by the allocation mechanism rules employed in an auction (such as the number of licences), the nature of the property rights offered (such as the duration of the licences to be auctioned) and by post-auction rules (such as whether trading of spectrum leases is permitted). Knowledge of these factors and expectations about the post-auction competitive market structure is used by bidders to develop business models which then determine their maximum bid value.

Maximum bid values may be interdependent, in that Bidder 1's value may be influenced by Bidder 2's value and depending upon allocation rules these could change during an auction. Not surprisingly in all auctions the valuations of bidders determine the outcome, and in a good spectrum auction, that is one that is well designed, spectrum should be assigned to those bidders holding the highest valuations.

5.2 Auctions versus administrative methods of assignment

The majority of spectrum licences in the world have been assigned via administrative means, though increasingly licences are being assigned via auction. In some cases, particularly for low value spectrum, administrative approaches are satisfactory. However, for high value spectrum, such as that used for mobile communications and broadcasting, the reliance upon bureaucratic machinery to decide on assignments is far from satisfactory.

A popular administrative method for assigning licences is the use of comparative hearings or *beauty contests*. Many radio administrations around the world continue to use this method for awarding high value cellular and other licences. In a large number of cases these award processes have given rise to legal challenge, as the scoring system used to determine winners is often alleged to lack objectivity and transparency. As Cave and Valletti [3] have stated "the beauty contest ... opens the door to favouritism and corruption".

Beauty contests do enable governments to award licences at low prices and arguably this could help promote the businesses of the licence holders. For example, Finland issued four 3G licences using a beauty contest in March 1999 with the intention of stimulating the development of UMTS service industries in Finland.[3] INTUG Europe [4], the international telecommunications user group, favours assignment by beauty contest because it is claimed they provide administrators with greater flexibility in the selection of operators and the outcomes of beauty contests lead to higher penetration rates in cellular markets, faster building of national networks and lower prices for users. However, these claims appear not to be supported by credible empirical evidence.

We view beauty contests for high value spectrum as a poor method for assigning spectrum rights. It should not be the job of radio administrators to second guess markets. As the Radiocommunications Agency [5] stated in preparing for the auction of 3G licences in the UK "Government should not be trying to judge who will be innovative and successful". In a statement on the pricing of 3G spectrum the Telecom Regulatory Authority of India [6] has stated that "the auction route is superior to the beauty contests and the fixed fee approaches". The latter

[3] Finland is where Nokia is headquartered.

Table 5.1. *Relative licensing assignment speeds*

Licensing mechanism	Licences issued	Time-line definition	Number of days
Comparative hearings	Cellular, non-wireline licences	Average number of days from application to grant of construction permit	720
Lotteries	Cellular, non-wireline licences	As above	412
Auctions	Broadband PCS licences	Average number of days from filing of short form application to licence grant	276
Auctions	All licences auctioned	As above	233

Source: FCC, 1997 The FCC Report to Congress on Spectrum Auctions.

view appears to be becoming more widely accepted among radio spectrum managers.

Another failing of the beauty contest approach is the lack of credibility – applicants tend to overstate service claims such as geographical and population coverage, service quality, etc. Indeed, applicants have little to fear from over-stating claims, especially if they know this will help them win a licence.[4] Another drawback to the comparative selection processes is the length of time and the amount of regulatory resources needed to conduct the exercise. In Table 5.1 it can be seen that comparative selection procedures to assign a licence on

[4] A similar problem can and has blighted some auctions where bidders have driven up prices and then defaulted on payment. However, suitable rules can be built into auctions to deter such outcomes.

average lasted 720 days in the USA against 276 days for a comparable licence assignment by auction. Even lotteries took longer to implement than the average auction.

5.3 Theory of auctions

Auctions typically involve either a large number of sellers and one buyer (often referred to as a reverse auction or procurement auction), or a large number of buyers and one seller. Radio spectrum auctions typically involve a large number of buyers (say phone companies, broadcasters, etc.) and one seller (the radio spectrum agency). Auctions can take many different formats and over the past 18 years or so a number of different auction formats have been used to assign radio spectrum licences.

In general there are two main forms of spectrum auctions: (i) interactive or open auctions and (ii) sealed bid auctions. Interactive auctions are where bidders interact with the auctioneer through a price discovery process that typically features either ascending or descending bids. An example of a popular interactive auction is the English auction in which bidders submit bids in an ascending manner until the number of bidders remaining equals the number of objects for sale. Another interactive auction is the Dutch auction, where the auctioneer starts with a high price and reduces this over time until a bidder accepts the posted price.

Sealed bid auctions are where bidders submit their bid only once in a sealed envelope (or more likely in practice via electronic means). The amount a successful bidder pays in a sealed bid auction depends on whether it is a first- or second-price auction. In a first-price auction successful bidders are required to pay the amount bid, whereas in a second-price auction the successful bidder pays an amount equal to the second highest bid submitted; see Vickrey [7].

Auction theory can be used to assess the efficiency properties of different auction formats, namely how well an auction leads to objects being obtained by those holding the highest values. The theory of auctions can also shed light on optimal bidding behaviour in auctions.

Under special conditions the amount of revenue an auctioneer should expect to receive in an auction has been shown by auction theorists to be invariant across a wide range of auction formats, see [7] and [8]. This result is known as the *revenue equivalence theorem* and has a significant bearing upon practical auction design and spectrum management policy.

If the revenue equivalence theorem holds for a wide class of settings, then it suggests that spectrum agencies need not worry too much about auction formats – as each format should on average achieve the same revenue. For example, it does not matter whether a first-price or second-price sealed bid auction is used, both will deliver the same revenue on average. However, practice has shown that auction design does have a significant bearing upon outcomes. Furthermore, auction theorists have identified a number of factors that undermine revenue equivalence across different auction formats. Nevertheless, the theorem is useful as it enables auction designers to identify those factors which may be responsible for affecting outcomes in an auction and hence it provides guidance on the optimal design of an auction.

In general, the key factors which influence expected revenues in an auction are related to the risk characteristics of the buyers and seller, and to the amount and type of information about the value of the objects held by bidders. In real world auctions there is likely to be much heterogeneity across bidders (e.g. incumbent telephone operators versus entrants) and this is likely to have a bearing on auction outcomes and on the design of an auction. In spectrum auctions with their mix of private and common values, the intensity of competition, for which the number of bidders is a proxy, generally has an impact on auction outcomes.

Auction theorists have shown for a wide class of auctions that a simple ascending auction with $N+1$ similarly placed (i.e. symmetric with respect to information) bidders will yield a better outcome than for the case where there are N bidders. This result is of special interest in the context of auctioning radio spectrum, as the number of bidders in high stakes auctions has been in many cases relatively small. Hence, the role of marketing an auction with the intention of attracting extra bidders will, at the margin, be a valuable exercise, see [9].

Auction theory is helpful in assisting spectrum agencies in deciding how best to package spectrum rights and on choosing the best auction format. Spectrum agencies need to determine the geographic scope of licences, the number of licences, the eligibility criteria for bidders and the precise auction rules.

5.4 Auction formats

The format of an auction depends on a number of factors, the most significant being the objectives of the spectrum custodian, which is usually government. If the custodian wishes to maximise revenue from the sale of spectrum, this may entail a different auction design than for the case where effective competition is the primary objective. Other important factors will be the quantum of spectrum available, the nature of demand-side interest in the spectrum, the geographic area covered and administrative factors. In many cases auctions for radio spectrum have been designed with efficient assignment in mind, though governments have also recognised that high value spectrum can be a useful way to collect extra tax revenue.

In the United States the Federal Communications Commission (FCC) and the National Telecommunications and Information Administration (NTIA) share responsibility for managing the spectrum. Section 309(j) of the Communications Act authorises the FCC to use auctions to promote efficient and intensive spectrum use as well as to promote the development and rapid deployment of new technologies, products and services for the benefit of the public, including those residing in rural areas, without administrative or judicial delays. The Act also requires the FCC to administer auctions so as to promote economic opportunity and competition, among other things. Other objectives in the USA include advancing spectrum reform by developing and implementing market-oriented allocation and assignment reform policies, and conducting effective and timely licensing activities that encourage efficient use of the spectrum, see [10]. However, some scholars have questioned whether the objective should be efficiency and argued instead that auctions are primarily a device for raising tax revenues (e.g. see Noam [11]).

The approach in Europe with regard to spectrum assignments has been to ensure harmonised conditions for the availability and efficient use of radio spectrum, see [12]. European licensing policy objectives require that where the demand for radio frequencies in a specific range exceeds their availability, appropriate and transparent procedures should be followed for the assignment of such frequencies, see [13]. In practice this amounts to the endorsement of assignment of valuable radio frequencies by auction.

In New Zealand, the first country to pioneer spectrum auctions, the government's policy objectives for spectrum encompass the promotion of competition; the maximising of the value of spectrum to society; and satisfying increasing demand. When assigning spectrum rights the government is required to adopt a market-based assignment process which is competitively neutral and transparent.

Spectrum auctions have been used in many other countries and in most cases the objectives are similar to those outlined above and can be summarised as follows.

- Efficiency: the assignment of licences leads to licences being awarded to those who value them the most (or put another way, awarded to those who contribute most to economic activity through using spectrum).
- Revenue: to raise tax revenue for the government.
- Competition: spectrum rights are issued in a way that helps promote effective competition.
- Transparency: to ensure that the process of selection is without corruption and undertaken expeditiously.

The challenge for spectrum agencies is to identify the appropriate auction formats for different types of radio spectrum uses and users that broadly achieve some or all of the above objectives. Below we discuss the most popular auction formats that have been applied by spectrum agencies around the world since 1989, as well as some formats that may be applied in the future.

5.4.1 Auction format options

(1) *First-price sealed bid auction* (equivalent to a Dutch auction). A very simple design that is easy to implement and easily understood by bidders, whereby the bidder who posts the highest bid wins the lot.

- Pros:
  - simple,
  - completes quickly,
  - ideal for small packages of spectrum and where each bidder will receive one licence in a given region,
  - good for competition – new entrants are more likely to enter and compete against incumbents.
- Cons:
  - bidders may shave bids to avoid the winner's curse, which in an uncertain environment could lead to much reduced bids,
  - bidders winning near identical objects may pay widely different amounts which could result in difficulties subsequently for managers or politicians,
  - no price discovery process,
  - reserve price needs to be carefully assessed.

There have been many spectrum auctions using the first-price sealed bid format. The first took place in New Zealand in October 1991 for FM, UHF, AM and DMS frequencies for local licences throughout the country.

In April 2006 the UK spectrum regulator Ofcom conducted its first spectrum auction and used this format. The auction was for low power services in the spectrum band 1781.7–1785 MHz, paired with 1876.7–1880 MHz (the GSM/DECT guard bands). These bands can be used for a range of applications, such as private GSM networks in office buildings or campuses, though formally the auction rules stated that the spectrum was to be awarded on a technology and application neutral basis. Ofcom used an innovative approach by allowing bidders to submit different bid values depending upon the number of licences that might finally be offered. Ofcom stated that it would issue a minimum of seven and a maximum of 12 licences, and that the actual number of licences would be determined by the auction process. This design feature allowed the market to determine how many operators should offer a service, rather than having the spectrum manager choose what it believed was best for customers.

In the auction 14 bids were submitted and the results are shown in Table 5.2. The outcome illustrates how different bidders can pay very different prices for the same object in a first-price auction. The bidder COLT paid a little over £1.5 m for a licence, whereas Spring Mobil AB paid £50 110.[5] The auction determined that 12 licences would be issued, as the bids submitted conditional on 12 licences being issued raised the highest revenue for Ofcom.

(2) *Second-price sealed bid auction* (equivalent to an English auction and also known as a *Vickrey* auction [7]) in which bidders post sealed bids and the highest bidder pays the price offered by the second highest bidder.

- Pros:
  - encourages bidders to reveal true valuations as paying according to the 'second valuation principle' safeguards against the winner's curse, leading to an efficient outcome – the Vickrey property,
  - ideal for small packages of spectrum and where each bidder will receive one licence in a given region,
  - completes quickly.
- Cons:
  - where the difference between the nth successful and the $n-1$th unsuccessful bidders' values is large, the winners may pay low amounts and this may be deemed politically unacceptable or embarrassing,

[5] The reserve price was £50 000.

Table 5.2. *Bids for the UK GSM/DECT guard bands*

	Number of licences on offer					
Bidder	**7**	**8**	**9**	**10**	**11**	**12**
BT	£305 112	£295 112	£275 112	£275 112	£275 112	£275 112
Cable and Wireless	£281 002	£281 002	£101 002	£51 002	£51 002	£51 002
COLT	£1 513 218	£1 513 218	£1 513 218	£1 513 218	£1513 218	£1 513 218
Cyberpress	£151 999	£151 999	£151 999	£151 999	£151 999	£151 999
FMS Solutions	£113 000	£113 000	£113 000	£113 000	£113 000	£113 000
Mapesbury Communications	£76 600	£76 600	£76 600	£76 600	£76 600	£76 600
O2	£209 888	£209 888	£209 888	£209 888	£209 888	£209 888
Opal Telecom	£155 555	£155 555	£155 555	£155 555	£155 555	£155 555
Orange	£50 000	£50 000	£50 000	£50 000	£50 000	£50 000
PLDT	£88 889	£88 889	£88 889	£88 889	£88 889	£88 889
Shyam Telecom	£101 011	£101 011	£101 011	£101 011	£101 011	£101 011
Teleware	£1 001 880	£1 001 880	£1 001 880	£1001880	£1 001 880	£1 001 880
Zynetix	£50 000	£50 000	£50 000	£50 000	£50 000	£50 000
Option Total	**£3 618 654**	**£3 721 654**	**£3 622 665**	**£3 687 212**	**£3 738 214**	**£3 788 324**

Source: Ofcom: (http://www.ofcom.org.uk/radiocomms/spectrumawards/completedawards/award_1781/notices/030506.pdf).

- no price discovery process,
- where there are many items to be auctioned and these comprise either substitutes and/or complements, this can give rise to an exposure problem (i.e. a bidder ends up with too many or too few licences), particularly if an auction is organised as a series of independent sub-auctions.

The first spectrum auction held in the world was a second-price sealed bid auction for UHF frequencies in New Zealand in December 1989. This was followed by two more second-price sealed bid auctions in 1990 in New Zealand. In the first auction the UHF frequencies result is shown in Table 5.3. One bidder, the Totalisor Agency, submitted bids NZ $401 000 for six licences, which then determined the prices paid for five of these licences.

As this auction comprised national and regional frequency rights where the lots offered were either substitutes or complements, it is doubtful whether efficiency was achieved. This is because the auctions were independent, which meant that it was difficult for bidders to

Table 5.3. *Result of the UHF auctions in New Zealand in 1989*

Near nationwide UHF-TV frequency rights			
	Successful bidder	Price paid	Highest bid
Lot 1	Sky Network Television	NZ $401 000	NZ $2 371 000
Lot 2	Sky Network Television	NZ $401 000	NZ $2 273 000
Lot 3	Sky Network Television	NZ $401 000	NZ $2 273 000
Lot 4	Broadcast Communications Ltd	NZ $200 000	NZ $255 124
Lot 5	Sky Network Television	NZ $401 000	NZ $1 121 000
Lot 6	Totalisor Agency Board	NZ $100 000	NZ $401 000
Lot 7	United Christian Broadcast	NZ $401 000	NZ $685 200

Source: (http://www.rsm.govt.nz/auctions/tender01/tender1.pdf).

manage the risk of winning too many or too few licences. A leading auction specialist Paul Milgrom [14] has stated that it is likely "the outcome was inefficient".

New Zealand abandoned second-price sealed bid auctions in 1990 and proceeded to use first-price sealed bids. This was largely in response to criticism levelled against the government for allowing large corporations to buy spectrum at knock-down prices, despite revealing in their bids that they would pay much higher prices. According to the Ministry of Economic Development of New Zealand [15] in one instance a bidder submitted a bid of NZ $100 000 and paid NZ $6 for a licence, and in other cases owing to the absence of a reserve price licences were given away because there was no second bid.

An innovative second-price sealed bid auction involving two different radio spectrum agencies, ComReg of Ireland and Ofcom of the UK, is scheduled to take place in 2007. The auction is for spectrum in the 1785–1805 MHz band, with one licence available in Northern Ireland and one licence in the Republic of Ireland. The coordination between the two agencies is intended to enable the strong complementarity between the two geographic areas to be realised.

(3) *Simultaneous ascending auctions* (SAA) are where bidders submit bids round by round, successively bidding up prices until a round is reached in which no new bids are received. The auctioneer stipulates a minimum bid increment and requires that bidders participate if they are active (this is known as the *Milgrom–Wilson* activity rule). Activity rules are intended to prevent the "snake in the grass" strategy, where a bidder may misrepresent early round bids by understating demand. Bidder identities can be kept confidential during the auction, to help eliminate collusion and signalling (this was a feature of the 3G auction in Hong Kong in 2001). The winners in an SAA should pay almost the same as in a second-price sealed bid auction. The amount paid will on average be slightly higher owing to the bid increment requirement.

- Pros:
  - efficient and exhibits Vickrey properties,
  - design can accommodate many spectrum blocks in one auction and blocks can be substitutes and complements,
  - different prices paid will reflect differences in values between objects in auction – arbitrage is possible across substitutes,
  - works effectively when competition is intense,
  - price discovery – especially significant where bidders seek to buy packages of licences,
  - activity rules can be designed to promote bidding with a view to accelerating completion of the assignment process.

- Cons:
  - may be time consuming,
  - implementation may be resource intensive,
  - demand reduction – "large" bidders may understate demand early on,
  - tacit collusion – bidders may seek to collude by submitting bids which act as signals (trailing digits have been used in some SAA auctions to send signals to other bidders in the hope of encouraging collusion), a small number of participants may encourage tacit collusion,

- exposure problem – if some spectrum licences are complementary a bidder may end up paying too much for some licences if the prices of complementary licences become too high in subsequent rounds,
- hold-up – opportunistic bidders might seek to target certain desirable packages,
- bidding strategies can be complex,
- vulnerable to collusion when competition is weak,
- agency problem – may lead to excessive prices if managers bid too aggressively fearing that an exit will be interpreted negatively by their principals,
- price discovery arguably of limited benefit in auctions where there are only a few licences available and where each successful bidder is restricted to one licence.

At the start of an SAA bidders are invited to submit bids on up to k spectrum blocks. For example, an SAA may feature five different areas each having two spectrum blocks making a total of ten spectrum blocks and where a bidder can at most acquire one licence in each area. In this example a bidder may be permitted to place bids on up to five spectrum blocks in the five different areas (hence $k = 5$). Suppose there are five bidders A, B, C, D and E.

Assume at the end of the first round that bidders A and C have submitted the highest bids on the two spectrum blocks in each of the five areas. In the next round bidders B, D and E are invited to bid (they are the eligible bidders), while bidders A and C will be ineligible to bid. Should one of the other bidders submit a higher bid than A or C, the auction proceeds to the next round, with A and/or C again eligible to bid. The SAA continues until a round is reached in which no new bids are submitted.

The conclusion of an SAA reveals the price bidders are willing to pay for a package of licences (where a package could comprise just a single licence), and the process of arriving at the revealed price is called the price discovery process. Auction specialists argue that the price discovery process is one of the key strengths of the SAA format.

In auctions where there are many related licences (usually meaning that there are licences in many different areas – though it could also refer to different types of frequency packages, such as paired and unpaired), the simultaneous sale of spectrum blocks and the ascending bid process aids price discovery which together enable bidders to assemble desirable packages of spectrum.

It is unquestionably the case that the SAA format has assisted the price discovery process in large scale auctions (those auctions involving many bidders and/or many licences) for awarding spectrum rights in the United States for which it was principally designed.[6] The SAA format was first deployed in the US in July 1994 for ten nationwide narrowband PCS (personal communications services: mobile voice and data) licences. The auction lasted five days and ended at round 47. Twenty nine bidders participated and at the end of the auction six bidders won ten licences raising $617 m.

Following the successful application of the SAA format for nationwide licences, the FCC then organised a series of SAAs for regional and local narrowband PCS licences. In these auctions many bidders were interested in acquiring several licences across different areas and the SAA price discovery property was critical to the success of the auctions.

In 1995 two blocks (A and B) of regional narrowband PCS spectrum generated net bid values just over $7 billion, and during 1995–96 the local narrowband PCS C Block auction raised over $10 billion in net bid values. The C Block auction lasted 184 rounds over the course of 84 bidding days. Two hundred and fifty five bidders qualified to participate in the C Block auction. At the end of the auction 89 bidders won 493 licences. At the time the scale of the auction was unprecedented and the sums raised were the largest ever seen for radio spectrum. These auctions were conducted electronically, which meant that the pace of the auction could proceed relatively quickly.

[6] Spectrum assignment in the USA is often undertaken on a regional and local level and many hundreds of licences are offered simultaneously.

The success of the SAA format in the US PCS auctions generated tremendous interest among radio spectrum managers around the world in auctions generally and in the SAA format in particular. To date SAA auctions have been used for many different types of radio spectrum globally and over $200 billion has been raised using this format.

New Zealand applied the SAA format to the assignment of local FM radio spectrum licences in October 1995 and September 1996 using a fax based bidding process. From 1998 onwards the New Zealand authorities used internet based bidding procedures to manage the SAAs – which enabled the auctions to be conducted more rapidly. Australia used the SAA format in 1998 to assign mobile telephony spectrum, yielding revenues on a per capita basis much higher than those raised in the US PCS auctions. In 1999 Canada applied the SAA format for spectrum in the 24 GHz and 38 GHz bands. By the late 1990s a number of European countries were interested in applying the SAA format for 3G spectrum.

In 2000 the SAA format was used to assign licences for 3G spectrum in a number of European countries. Unlike the application of the SAA format in the US, the 3G licence assignments in Europe were relatively small scale affairs involving usually no more than six nationwide licences and where the number of qualified bidders in most cases was also relatively small. Table 5.4 provides data on the number of licences available in the main European 3G auctions using the SAA format held over the period 2000–01.

It is debatable whether the SAA format added much value to efficient assignment of spectrum in the European 3G auctions. Bidders in these auctions were not assembling packages of different licences (except sequentially across different auctions in different countries), the number of licences in each auction did not exceed six, the number of bidders was small, and the bidders were arguably well informed sophisticated entities who had sufficient time to prepare and assess the value of the licences. In these circumstances the SAA was vulnerable to both collusion and the agency problem, and it is unclear what would be gained from price discovery. A suitably designed second-price sealed bid auction format would have been simpler, quicker and

Table 5.4. *3G SAA auctions in Europe 2000–01*

Country	Date	Number of bidders	Number of licences	Comments
United Kingdom	April 2000	4 incumbents, 9 new entrants	5	Auction extended over 52 days and 150 rounds
Netherlands	July 2000	5 incumbents, 1 new entrant	5	Auction extended over 14 days and 305 rounds
Germany	August 2000	4 incumbents, 3 new entrants	6	Auction extended over 19 days and 173 rounds
Italy	October 2000	4 incumbents, 2 new entrants	5	Auction extended over 2 days and 11 rounds
Austria	November 2000	6 bidders	6	Auction extended over 2 days and 14 rounds
Switzerland	December 2000	4 bidders	4	Auction lasted 1 day
Belgium	March 2001	3 incumbents	4	Auction extended over 2 days and 14 rounds

Source: NAO: HC 233 Session 2001–2002: 19 October 2001.

arguably more effective at revealing bidders' valuations and less vulnerable to collusion and the agency problem.

The 3G auction in the UK in 2000 resulted in the largest revenue ever raised from a spectrum sale when measured on a per capita basis. For each person in the UK £375 ($694 at an exchange rate $1.85 = £1) was raised in revenue, which contrasts with the PCS C Block auction where revenue per person was much lower at a little over $38.[7]

[7] The US population at the time of the auction was 262 million, see [16].

Table 5.5. *Results of the UK 3G SAA*

Licence	Winning bids (£bn)	Price per MHz (£m)	Successful bidder
A	4.385	125	TIW
B	5.964	199	Vodafone
C	4.030	161	BT Cellnet
D	4.004	160	one2one
E	4.095	164	Orange
Total	**22.478**		

Source: NAO: HC 233 Session 2001–2002: 19 October 2001.

The UK 3G auction offered paired spectrum, which is especially valued for mobile communications applications, and unpaired spectrum, which was valued less. Five national 3G licences were made available by the spectrum agency in the UK, with Licence A comprising 15 MHz paired and 5 MHz unpaired spectrum reserved exclusively for new entrants. Of the other four lots, Licences, C, D and E were identical in size at 10 MHz paired and 5 MHz unpaired (though differed in spectral location) whereas Licence B was larger at 15 MHz paired. Thirteen bidders competed in the auction, of which four were incumbent cellular phone network operators.

The result of the UK 3G auction is shown in Table 5.5. The auction ended after 150 rounds on 27 April, lasting seven weeks. The successful bidders were the four incumbent operators and a new entrant TIW, who subsequently sold the A Licence to Hutchison Whampoa. On a per MHz basis, the prices paid for Licences C, D and E reflect the arbitrage property of SAA.

Table 5.6 illustrates the highest bids submitted by each of the bidders in the UK 3G auction. It can be seen that five bidders exited when the bid values were around the £2 billion level per licence, which occurred between rounds 90 and 97. After this the auction continued with eight bidders until round 131 when WorldCom exited the auction, and thereafter bid values increased to over £4 billion per

Table 5.6. *Bidding process in the UK 3G auction*

Bidder	Round	Licence	Bid value (£bn)
Crescent	90	C	1.819
3G	90	A	2.001
Epsilon	94	C	2.072
Spectrum	95	D	2.100
OneTel	97	E	2.181
WorldCom	119	D	3.173
Telefonica	131	C	3.668
NTL	148	C	3.971

Source: NAO: HC 233 Session 2001–2002: 19 October 2001.

licence. The last unsuccessful bidder in the auction was NTL, which submitted its last bid value £3.97 billion in round 148 on Licence C. Earlier in the auction in round 127 NTL had submitted a higher bid of £4.28 billion on the larger Licence A, though this was equivalent to £143 million per MHz versus £159 million per MHz in its final bid for Licence C.

Similar SAA auctions for 3G spectrum were held in a number of other European countries and the outcomes are illustrated in Figure 5.2. The revenues received vary substantially, with later auctions doing particularly badly. The auction held in Switzerland was a notable disaster, as the number of bidders ended up equal to the number of licences, leading to weak competition and collusive behaviour as bidders exited or formed larger coalitions prior to the start of the auction.

The failure of 3G auctions that were held in late 2000 coincided with changing market sentiments at the time, with the stock market crashing and the value of telecommunications companies in particular being adversely affected. This was the time of the so-called dotcom crash. In the UK, which had experienced the most successful SAA ever, only a few months later in November 2000 an SAA for broadband fixed wireless access (BFWA) spectrum in the 28 GHz band was a notable failure. The failure of the UK BFWA auction was ironic as the SAA

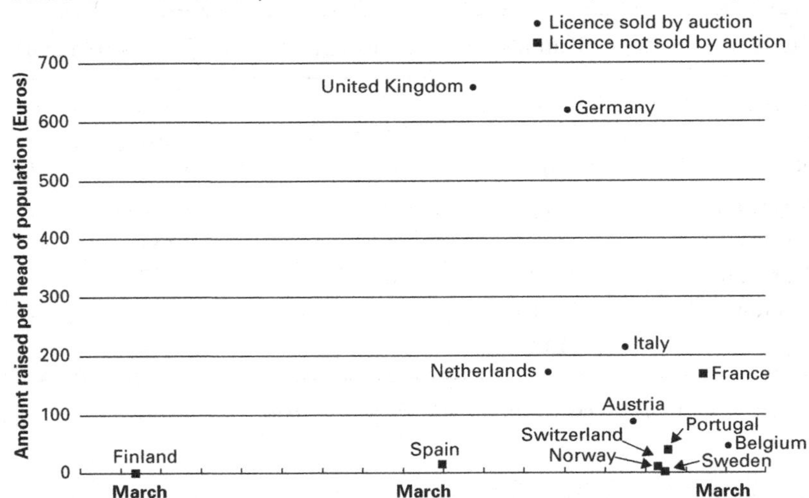

Figure 5.2. 3G auction bid values around Europe. Source: NAO. HC 233 Session 2001–2002: 19 October 2001.

format was better suited for this spectrum assignment process than the 3G assignment.

The UK BFWA auction was conducted on a regional basis with three licences available in each of fourteen regions. Unlike the UK 3G auction where a bidder could purchase only a single national licence, in this auction a bidder could purchase anywhere between one and fourteen licences across the different regions (each bidder was restricted to one licence in each region). The auction featured substitutes (the three licences in each region) and complements (licences in the different regions) and an SAA format was by far the most suitable format.

At the end of the auction, however, only sixteen licences in eight regions were sold out of the possible 42 licences. The total proceeds amounted to £38.16 million, compared to predicted proceeds of over £1 billion. The BFWA auction failed for a number of reasons – though none is connected with the SAA format. The most significant factor was an adverse change in market conditions in the telecommunications sector during the latter half of 2000. The changing market conditions did not get reflected in the reserve prices, as these were established on rather favourable views prior to the market meltdown. Hence, bidders decided not to submit bids on available licences because the reserve price was too high. Furthermore, no procedure was set in place for dealing with the unsold licences.

The setting of reserve prices is often an important feature of spectrum auctions, though it is not always necessary.[8] The purpose of a reserve price is to reflect the value of spectrum to "society" and to ensure that it is not transferred at knock-down prices below the true economic value.[9] Indeed, if reserve prices are set too low, this could encourage unhealthy tacit collusion as bidders may benefit significantly from an early conclusion to an auction. This may have been one reason contributing to the failure of the Swiss 3G auction in 2001, when the licences were eventually assigned at values (assessed on a per capita basis) significantly below those in comparable markets such as Germany and the UK.

On the other hand, if reserve prices are set too high this can reduce participation, which is bad for competition and the price discovery process. In the UK BFWA auction the reserve price was based on data that were out of date by the time the auction occurred. The setting of reserve prices is challenging and requires careful judgement about the market prospects. For radio administrators setting reserve prices, erring on the side of caution and being slightly pessimistic is probably advisable, though being overly pessimistic can give rise to embarrassments as in the Swiss 3G auction.

More recently an SAA format has been used to assign major new spectrum in the United States. In 2006 an SAA auction for advanced wireless services (AWS), an array of innovative wireless services and technologies, including voice, data, video, and other wireless broadband

[8] The Dutch 3G auction did not feature reserve prices and three of the six bidders in the auction submitted opening bids of zero. The auction finally concluded after 305 rounds after one of the bidders (Telfort) sent another bidder (Versatel) a letter threatening a law suit if bidding continued!

[9] The economic value of spectrum is its opportunity cost – a concept which is discussed in detail in Chapter 11 on administrative incentive prices for radio spectrum.

services offered over Third Generation ("3G") mobile networks, was held. The AWS auction, like previous such auctions in the United States, illustrates that the SAA format is very scalable and can cope with large numbers of bidders and licences. The FCC organised the AWS auction (it was FCC auction number 66) in which there were 1122 licences available: 36 Regional Economic Area Grouping (REAG) licences, 352 Economic Area (EA) licences, and 734 Cellular Market Area (CMA) licences; see Figures 5.3 and 5.4 for the spectrum and areas defined in the auction.

In most previous spectrum auctions organised by the FCC a patchwork of licences across the USA covering as many as 714 areas or at a minimum 176 licences had been the norm. The approaches in the past were taken partly to encourage the development of small rural providers, but it has meant that not one single wireless provider covers 100% of mainland USA. In the AWS auction the mainland USA could be covered by winning only six regional licences.

The FCC auction of AWS spectrum licences started on 9 August 2006 and ended on 18 September 2006 in round 161. Of the 1122 licences offered, 104 bidders won 1087 licences, raising a total of $13.9 billion. The AWS auction was the 64th auction completed by the FCC in 13 years since it was granted competitive bidding authority.

The FCC is planning further SAA auctions and is debating whether to modify the format to allow for package bidding. This is discussed further in point (7) below in the section on the combinatorial auction format.

(4) *Ascending clock auction (also known as a Japanese auction or button auction)*[10] in which bidders submit their demand in response to announced prices issued by the auctioneer. The auction continues until

[10] A variant of this auction is known as the *Ausubel Auction* [17] in which bidders reveal demand to announced prices. Items are awarded at the current price whenever they are "clinched" and the price is incremented until the market clears. It has been shown that this auction has the same efficiency properties as a Vickrey auction, and in circumstances where bidders' values may be interdependent (which is likely in some spectrum auctions) is generally superior to the Vickrey auction.

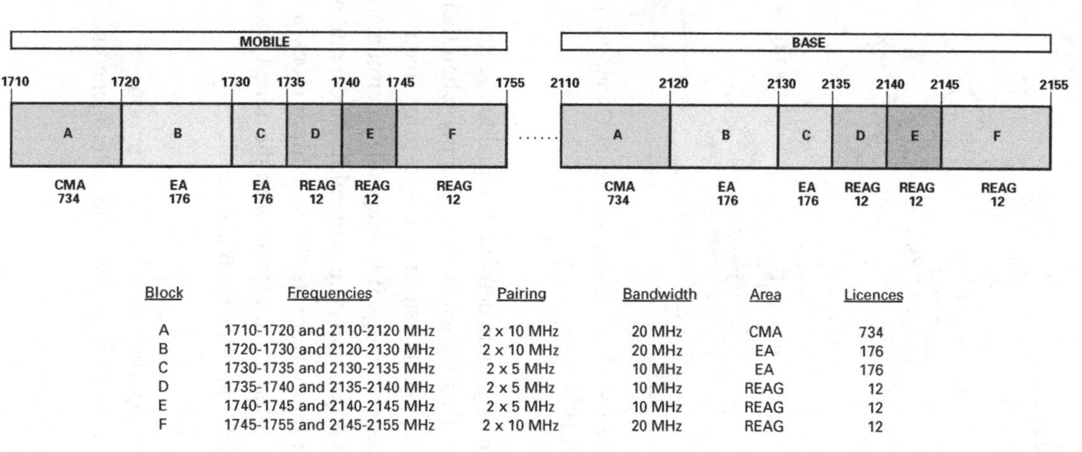

Block	Frequencies	Pairing	Bandwidth	Area	Licences
A	1710-1720 and 2110-2120 MHz	2 x 10 MHz	20 MHz	CMA	734
B	1720-1730 and 2120-2130 MHz	2 x 10 MHz	20 MHz	EA	176
C	1730-1735 and 2130-2135 MHz	2 x 5 MHz	10 MHz	EA	176
D	1735-1740 and 2135-2140 MHz	2 x 5 MHz	10 MHz	REAG	12
E	1740-1745 and 2140-2145 MHz	2 x 5 MHz	10 MHz	REAG	12
F	1745-1755 and 2145-2155 MHz	2 x 10 MHz	20 MHz	REAG	12

Figure 5.3. Spectrum in the US AWS auction. Source: FCC (http://wireless.fcc.gov/services/aws/data/awsbandplan.pdf).

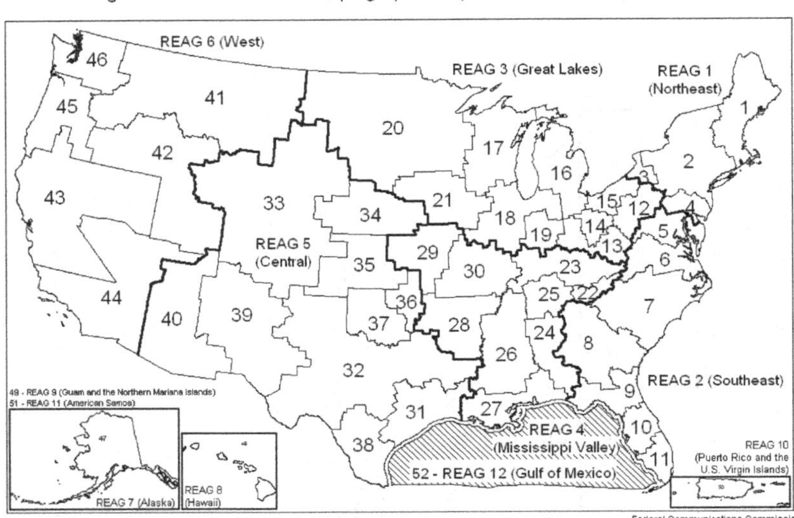

Figure 5.4. Areas defined in the US AWS auction. Source: FCC (http://wireless.fcc.gov/auctions/data/maps/reag.pdf).

a round is reached in which aggregate demand is equal to the number of spectrum licences on offer. This format is easy to apply when there are identical (or near identical objects) substitute spectrum licences and bidders are seeking one licence, though it is adaptable to circumstances where bidders may seek multiple units (i.e. several licences) and where items may be different. In many respects the ascending clock auction has very similar properties to an SAA auction with the added attraction that it is more likely to yield an efficient outcome.[11]

- Pros:
  - efficient and exhibits properties of a Vickrey auction,
  - works well for auctions with substitute and heterogeneous packages,

[11] An auction variant similar to the ascending clock auction is a *survival auction*. This is a multiple round sealed bid auction in which the lowest bids are successively eliminated and which form the reserve price in the subsequent round. This auction format is very similar to the ascending clock auction format and has similar properties. To date we are not aware of this auction format having been used for the assignment of radio spectrum.

  - simple,
  - transparent,
  - auctioneer can determine pace of auction and the announcements combat collusive bid signalling,
  - good for multi-unit auctions.

- Cons:

  - may require two auctions (sequential, not simultaneous) if the auctioneer overshoots the valuations of the bidders,
  - may require sequential auctions where packages differ.

The world's first ascending clock auction for radio spectrum was successfully deployed in Nigeria in 2001. In this auction three identical GSM licences (other than with respect to spectral location) were offered and there were five bidders. This auction also incorporated a sealed bid phase in the event of over-shooting by the auctioneer. The design was similar to the Anglo-Dutch auction format, see point (9) below.

(5) *Revenue share auctions* are variants of standard auction formats in which bidders submit royalty bids based on the share of revenue they will receive.

- Pros:

  - said to share market risks between auctioneer and bidders,
  - simple.

- Cons:

  - the royalty spectrum licence fee forms part of a recurring cost base, which means market prices incorporate a tax.

The most significant spectrum revenue share auction to date occurred in Hong Kong in 2001 for 3G licences. The auction was designed allowing participants to bid royalties with a reserve price of a 5% royalty on network revenues. The auction was a second-price auction, as the royalty figure was determined by the highest unsuccessful bidder.

(6) *Hybrid auctions*, which combine auctions with beauty contests (combinations of auction formats, which are also hybrids, are considered separately below). The process of comparative selection usually

takes place at the pre-qualification stage to determine which bidders become eligible to participate in the auction.

- Pros:
    - agency has greater discretion over qualification of bidders.
- Cons:
    - is susceptible to the deficiencies of a pure comparative selection process.

This approach was adopted during the registration and pre-qualification stage of Hong Kong's 3G auction.

(7) *Combinatorial auctions* are suitable for multi-lot auctions where the objects have potentially strong complementarities and where bidders' valuations may vary markedly. In a combinatorial auction bidders may submit package bids, i.e. bid a single price on a package of frequency rights.

- Pros:
    - can work with most auction formats, though a sealed bid approach is much easier to implement,
    - can result in substantial lowering of the exposure problem,
    - suited to auctions with a relatively small number of licences that are complementary.
- Cons:
    - complicated and may be difficult for bidders and observers to understand, which may undermine participation and revenue,
    - complex to implement,
    - requires detailed market understanding to assess what packages are most desirable,
    - may require sequential implementation to reduce complexity of auction,
    - little tested in practice,
    - transparency sacrificed when viewed from a public accountability perspective.

The first combinatorial spectrum auction occurred in Nigeria for fixed wireless access licences in June 2002. Eighty licences were offered across 37 regions, with two or three licences available in each region. In principle, a bidder might have been interested in submitting bids for all the packages containing five licences. The total number of five licence packages across the 37 regions is 52 307 640! It is inconceivable to envisage an auction in which bidders submit bids on possibly hundreds of millions of different packages.

The logistics of the Nigerian auction were simplified by the auction designers, who elicited from the bidders those regions where interest was greatest. This led to a premium set of regions being made available first, followed by licences in smaller regions. In all five sealed bid combinatorial auctions were held in sequence, with Auction 1 followed by Auctions 2 and 3 held in parallel and Auctions 4 and 5 held in parallel.

In 2006 Ofcom in the UK started preparations for organising its first combinatorial auction for four lots of paired frequencies in the 400 MHz band: lots A, B, C and D. Bidders are invited to submit at least one bid, but can place up to 15 bids on all the possible combinations using a sealed bid. The winners will be those submitting the highest bids and those bids which when combined yield the highest revenue. In a presentation made by Ofcom on 5 September 2006 an illustration of how the winners are determined was given and is reproduced in Table 5.7.

The winning bids are determined by searching for the combination of bids that produces the highest total value. The highest total value combinations of bids across the four bidders are as follows.

- 1st BC (Dick, £240) and AD (Emma, £180) = £420
- 2nd BC (Dick, £240), A (Tom, £80) and D (Emma, £90) = £410

The winning bids are for Licences BC and AD and the winning bidders are Dick (for Licence BC) and Emma (for Licence AD), as indicated in bold font in Table 5.7.

(8) *Sequential auctions* are auctions where spectrum licences are offered in sequence. For example, if a spectrum agency is offering

Table 5.7. *Determining winners in the 400 MHz band auction*

Licences	Tom	Dick	Harry	Emma
A	£80	£80	—	£90
B	£80	£100	—	£90
C	£70	£100	—	£90
D	£70	£70	—	£90
AB	£170	£200	—	£200
AC	£160	—	—	£180
AD	£160	—	—	**£180**
BC	£170	**£240**	—	£200
BD	£160	—	—	£180
CD	£170	£180	—	£200
ABC	—	—	£290	£300
ACD	—	—	—	£280
BCD	—	—	£290	£300
ABCD	—	—	£380	£390

several identical licences, it might decide to offer licence 1 in auction 1, followed by licence 2 in auction 2, and so on.

- Pros:
  - can work with most auction formats,
  - easy to administer.
- Cons:
  - encourages strategising among bidders and may give rise to peculiar outcomes.

National and regional BFWA licences were auctioned sequentially in Switzerland in March 2000. Two national licences were auctioned, the first on 8 March and the other on 9 March. Four bidders participated in the auction on 8 March and the successful bidder UPC paid SFR120.85 million, outbidding FirstMark Communications who bid SFR115.09 million, see Tables 5.8 and 5.9.

In the auction held the next day, the three unsuccessful bidders from the day before participated. The outcome resulted in FirstMark Communications winning and paying SFR134 million, some SFR14 million more than it bid in the previous day. The outcome is peculiar as FirstMark chose to stop bidding in the first auction at SFR115 million. Nevertheless, the outcome of the auction is consistent with the theoretical literature on auctions, which suggests that prices may drift upwards in sequential auctions if there is a common value element.

Turkey held a sequential auction in 2000 for two mobile telephony spectrum licences in the 1800 MHz band, employing a rule that the reserve price in the second auction would equal the price paid in the first auction. This design was flawed as it encouraged strategic excessive bidding in the first auction which led to a high reserve price in the second auction such that no bidder was prepared to pay. Thus the successful bidder in the first auction (Is-TIM) strategically generated a more concentrated market structure as a result of the poor auction design. Furthermore, valuable spectrum was left idle and the government lost revenue.

(9) *Anglo-Dutch auction – an SAA auction followed by a sealed bid auction.*

- Pros:
  - has the same merits as an SAA auction,
  - can be designed to accommodate subsequent auction of unsold frequencies in cases where the reserve price is misjudged.
- Cons:
  - as for SAA and first-price sealed bid auctions above.

This auction has properties similar to an English auction, but a first-price sealed bid element at the end makes the conclusion akin to a Dutch auction. The pure Anglo-Dutch auction involves an ascending auction which ends when the number of bidders is one greater than the number of spectrum licences available. Thus, had this format been used

Table 5.8. *First round in the Swiss BFWA auctions*

Price (SFR)	Time	Company	Bid increment (%)
120 849 829	14:30	United	5.0
115 095 075	14:24	FirstMark	5.0
109 614 357	14:14	United	5.0
104 394 626	14:05	FirstMark	5.0
99 423 453	13:55	United	5.0
94 689 003	13:46	FirstMark	5.0
90 180 003	13:36	Europe i	12.9
79 864 866	13:32	Callino	5.0
76 061 777	13:23	Europe i	5.0
72 439 266	13:20	Callino	5.0
68 989 777	13:15	Europe i	5.0
65 704 525	13:10	Callino	5.0
62 575 738	13:07	FirstMark	5.0
59 595 941	12:58	Callino	5.0
56 758 039	12:55	FirstMark	5.0
54 055 275	12:46	Callino	5.0
51 481 214	12:41	FirstMark	5.0
49 029 728	12:32	Callino	5.0
46 694 979	12:30	FirstMark	5.0
44 471 409	12:23	Callino	5.0

Source: http://www.bakom.ch/ (Swiss telecoms regulator).

to assign the UK 3G licences, the SAA part would have stopped after round 131 when the Spanish operator Telefonica exited the auction leaving six bidders chasing five licences. At this point in the auction the bid values for licences C, D and E were just under £3.7 billion, and there would then have been first-price sealed bids submitted for each of the licences where the reserve price would have been close to £3.7 billion.

A variant of this format, though strictly not a pure Anglo-Dutch auction, was used to issue spectrum licences for public fixed wireless access (PFWA) broadband internet services in the 3.4 GHz band in the UK in 2003. Fifteen PFWA licences, one for each defined geographic

Table 5.9. *Second round in the Swiss BFWA auctions*

Price (SFR)	Time	Company	Bid increment (%)
134 009 564	13:22	FirstMark	5.0
127 628 156	13:13	Callino	5.0
121 550 625	13:03	FirstMark	5.0
115 762 500	12:54	Callino	5.0
110 250 000	12:45	FirstMark	5.0
105 000 000	12:36	Callino	5.0
100 000 000	12:26	FirstMark	10.2
90 720 007	12:17	Europe i	12.0
80 999 100	12:16	FirstMark	5.0
77 142 000	12:07	Europe i	5.0
73 468 571	12:05	Callino	5.0
69 970 068	11:56	Europe i	8.5
64 507 000	11:54	FirstMark	5.0
61 434 567	11:53	Europe i	5.0

Source: http://www.bakom.ch/ (Swiss telecoms regulator).

region of the UK, were available in the auction and 26 consortia applied to participate. The sealed bid element was planned to take place if one or more licences remained unsold after the SAA stage had completed. The auction started on 6 June 2004 and ended after 41 rounds on 17 June and lasted a total of 8 business days. The auction did not enter the sealed bid phase as all 15 licences were sold to three bidders.

(10) *Clock-proxy auction.* This auction format has been discussed widely among spectrum auction specialists in recent years and has been considered for implementation by the FCC. The auction comprises a first phase which is an ascending clock auction, the result of which would send valuable price signals to the participants which form the initial minimum bids in the proxy phase.

- Pros:
  - clock element is good for price discovery,
  - the proxy element can eliminate the exposure problem,

    - proxy phase achieves efficient assignment,
    - proxy phase has no incentives for demand reduction,
    - proxy phase is good for deterring collusion.

- Cons:

    - untested,
    - unclear whether bidders would reveal true valuations in an actual proxy phase.

As the announced prices in the clock auction are linear, this means that packages of licences may not be assigned as efficiently as they would if prices were non-linear.[12] The second phase is an auction conducted by proxy agents. The bidders report their valuations to proxy agents for the different packages they are interested in acquiring. The proxy agent then bids in an ascending package auction on behalf of the real bidder, aiming to maximise the bidder's payoff (difference between the amount paid for the package and the value of the package). When the proxy agents submit bids, the seller chooses those bids which maximise revenue.

In effect the proxy phase is akin to a sealed bid auction where bidders submit valuations, the clock phase prices form the initial bids and a computer programme then determines the successful bidders. Although it is vastly more complex than the Anglo-Dutch auction discussed above, it shares many similarities. Neither of the formats have to date been implemented.

5.5 Auction logistics

An auction is one part of a four stage assignment process: an invitation stage; a pre-qualification stage; an auction stage; and a grant stage. These stages follow a sequence, from promoting awareness of the

[12] A linear price is one where the unit price of spectrum is given irrespective of the amount of spectrum demanded by a bidder. A non-linear price is one where the unit price changes as the amount demanded changes. Bulk discounts are a good example of non-linear prices. Economists have demonstrated that in many circumstances non-linear price schemes can perform better than simple linear price schemes.

auction among potential bidders (marketing) to the eventual assignment of licences to successful bidders (the grant stage). This section describes in greater detail each of the stages involved.

5.5.1 The invitation stage

The invitation stage usually comprises the publication of an Information Memorandum, containing details about all subsequent stages of the auction, application forms, and pre-qualification requirements. This stage is an important part of the overall process as it includes the period in which to market the auction. If revenue is an important criterion, then participation will be important.

5.5.2 Pre-qualification stage

The pre-qualification stage in an auction provides an opportunity for the spectrum agency to screen out "inappropriate" bidders and to learn more about likely demand. This is achieved by setting out criteria that must be satisfied by all prospective "appropriate" bidders before being accepted as participants. In effect a pre-qualification stage is a hurdle to participation. No comparison is made across bidders, distinguishing it from a beauty contest or comparative selection process.

The criteria used should ideally be objective, transparent, and not impose a significant burden on prospective bidders. After all, one of the main purposes of an auction is to delegate assignment to competitive forces rather than bureaucracy. For example, in the Nigerian GSM auction the pre-qualification criteria comprised a few straightforward questions selected principally to:

- deter those engaged in money laundering (a major concern in Nigeria) or other illegal activities from participating;
- ensure that bidders satisfied ownership rules designed to stop bidder collusion; and
- ensure that bidders had some experience of delivering telephony services.

As part of pre-qualification it is usual in spectrum auctions to require would-be bidders to deposit a sum of money with the auctioneer. By requesting up-front deposits, financial penalties for breach of auction rules can be made more credible. Deposits are also used by auctioneers as a part-payment towards the objects won by successful bidders, and to this end can reduce the risk and offset the costs associated with default.

Deposits, and more generally the payment terms for an object won in an auction, affect the degree of speculative activity that may occur in an auction. Speculation tends to arise in situations where information is imperfect, and particularly where some bidders are better informed than others – the case of asymmetric information. It can also occur where the rules of an auction are biased towards favouring participation by certain types of bidder, giving rise to asymmetry among participants.

Speculative bidders seek to win objects solely to sell on subsequently at, hopefully, higher prices. As the market price of an object won in an auction is not typically known with certainty prior to an auction, speculators may default on paying for an object won in an auction if it is discovered at the conclusion of an auction that the market price is below the auction price. The level chosen for the deposit can help reduce speculative activity, and is therefore an important instrument available to auction designers.

A simple example helps to illustrate this point. Suppose there are two possible market prices for an object at the conclusion of an auction, MH and ML, where $MH > ML$. Suppose the auction price for the object is A, where $MH > A > ML$. (Note a bidder at the conclusion of the auction would pay A less the deposit D, the latter having already been paid to the auctioneer.) Assume that market price MH occurs with probability p, and market price ML occurs with probability $1-p$. Finally, let D be the deposit required to become an eligible bidder in the auction, which would be forfeited if the bidder were to fail to pay at the conclusion of the auction.

A speculative bidder is interested only in acquiring the object in the auction for subsequent resale at a profit. A speculative bidder would default on payment if the market price were ML, resulting in a loss equal to the deposit D. If the market price were MH, a speculative bidder would obtain a positive return equal to $MH-A$. The overall expected return to the speculative bidder can be expressed as follows:

$$p[MH - A] + (1 - p)(-D).$$

If the expression above were positive, a risk neutral speculative bidder would participate in an auction. The deposit D can be set at a level whereby the expression is equal to zero or even negative. Of course, the higher the value of D, the lower the expected return from speculative bidding.

In practice an auctioneer operates in a far more complex and uncertain setting than is suggested in this example. Nevertheless, it usually remains the case that a higher deposit provides a greater deterrence to speculative bidding.

As a further deterrent to speculative activity, it is common practice to require the successful bidder to pay in full the licence fee within a given period following the auction closure. If a successful bidder defaults on payment, various measures can be taken such as barring entities from any future licence assignments for a period of years and forfeiting the deposit.

If the risk of default is a concern in a spectrum auction, other measures can reduce it further. One common instrument that has been used in a number of spectrum auctions is the requirement that bidders submit bank guarantees before, and even during, an auction. This requirement places the risk assessment burden onto banks. As banks are better informed about the risk characteristics of bidders, the probability of default is likely to be lower in the presence of bank guarantees. The insistence on bank guarantees is sometimes welcomed by "serious" bidders, as it means that the auction price is more likely to reflect fundamentals and will not be inflated by insincere bidding.

Bank guarantees were not used to any significant extent in the European 3G auctions, as default was not regarded to be a problem. In advanced economies with mature capital markets, shareholder governance will usually ensure sincere bidding in high stakes auctions. In economies with less sophisticated capital markets, monitoring of management by shareholders is likely to be less effective at preventing insincere bidding. In such circumstances there may be good reasons to insist upon bank guarantees.

5.5.3 Auction stage: designing the rules

Auction rules are often very detailed and cover activities by bidders before, during and immediately after the auction. Rules are required to prevent collusive behaviour that could undermine efficiency, and to provide detailed guidance about what is and what is not permitted during an auction. Detailed and precise rules are required so as to give bidders as much certainty as possible, encouraging them to concentrate on valuing licences rather than engaging in excessive strategising against each other. Spectrum auction rules are typically drafted by economists and lawyers, who together work with the auctioneer.

Auction designers often specify rules for almost all eventualities, even for circumstances that may seem highly unlikely. Detailed rules are required because should an unlikely event occur and no rules exist, the process falls into disarray, as rules have to be made up on the spot. Furthermore, it is vital that bidders know precisely what the consequences will be for any action they take. Otherwise the uncertainties can cause them to do "bad" things either intentionally or unintentionally. For example, if there were some action a bidder could take whose consequences were not pre-specified, they could attempt to game the system by doing that and throwing the auction process into the courts to be resolved. This has been attempted numerous times in FCC auctions (though with little success).

In many SAA spectrum auctions, bidders are given an opportunity to pause and reflect on their strategy. This is usually permitted in two ways: via "waivers" and through the calling of a "recess". The action of a waiver is where a bidder is allowed to abstain from making a bid in a round. A waiver is intended to give a bid team pause for thought and possibly time to communicate with financiers and other interested parties.

Whether or not to allow waivers in an auction poses a challenge for the auction designer. Bidders expect and should be granted a reasonable amount of time to discuss their valuation, particularly as bidders in high stakes spectrum auctions are typically consortia whose members may have different views and means of access to capital. However, allowing bidders time to compose bids by granting waivers means there is a chance they could use waivers to mislead other bidders, or to take advantage of information revealed in an auction by other bidders. In the UK 3G auction bidders were permitted a maximum of three waivers and one recess day.

Bid increments need to be determined by the auctioneer and can be chosen to accelerate the pace of an auction if bidding is too slow.

5.5.4 The grant stage

The successful bidders in an auction progress into the grant stage, when licences are issued subject to receipt of monies owed to the auctioneer.

5.6 Conclusion

Spectrum auctions have been used extensively around the world for assigning many thousands of licences covering many different uses and types of user and to date are the most transparent and successful of the market methods applied in spectrum management. Some auctions have raised vast sums of money, while other auctions have raised very little. Competition has varied in different auctions, though the degree of competition often reflects broader market sentiments.

Since the first spectrum auction in 1989, the design of auctions has evolved to accommodate very large scale auctions and small scale auctions. The SAA format is often used for large scale auctions and is particularly good where licences may be similar though not identical and where there are complementarities.

Following the application of the SAA format for assigning 3G spectrum licences in Europe in 2000 and 2001, radio administrators in Europe have moved away from this format and made greater use of the simpler sealed bid formats. Indeed for small scale auctions where there is little package bidding, sealed bid auctions would seem the most appropriate.

More recently, auction theorists have been debating the merits of more sophisticated auction formats that combine the merits of both sealed bid and ascending bid auction formats. Two such formats have been discussed extensively in the academic literature and considered for application in spectrum auctions. These are the Anglo-Dutch auction and the clock-proxy auction. It is likely that one or both of these formats will be used in a future spectrum auction.

References

[1] "The FCC report to congress on spectrum auctions", Federal Communications Commission, Wireless Telecommunications Bureau WT Docket No. 97–150, 30 September 1997, available at http://wireless.fcc.gov/auctions/data/papersAndStudies/fc970353.pdf

[2] E. Kwerel, "Spectrum Auctions Do Not Raise the Price of Wireless Services: Theory and Evidence", Office of Plans and Policy, Federal Communications Commission, October 2000.

[3] M. Cave and T. Valletti, "Are spectrum auctions ruining our grandchildren's future?", *Info*, **2**(4), 347–50, 2000.

[4] INTUG Europe, "Licensing 3G in Europe: the users' view" available at http://www.intug.net/talks/ES_access_2000_12/text.html, 2000.

[5] Radiocommunications Agency, "Auction of Third Generation Mobile Telecommunications Licences in the UK: frequently asked questions", February 1999, available at http://www.ofcom.org.uk/static/archive/spectrumauctions/documents/faq2.htm#The%20Auction

[6] Telecom Regulatory Authority of India, "Recommendations on allocation and pricing of spectrum for 3G and broadband wireless access services", 27 September 2006, New Delhi.

[7] W. Vickrey, "Counterspeculation, auctions and competitive sealed tenders", *Journal of Finance*, **16**, 8–37, 1961.

[8] P. Klemperer, *Auctions: Theory and Practice*, Princeton University Press, 2004.

[9] J. Bulow and P. Klemperer, "Auctions vs. Negotiations", *American Economic Review*, **86**, 180–194, 1996.

[10] See http://www.fcc.gov/spectrum/

[11] E. Noam, "Spectrum Auctions: Yesterday's Heresy, Today's Orthodoxy, Tomorrow's Anachronism. Taking the Next Step to Open Spectrum Access", *Journal of Law and Economics*, December 1998.

[12] Article 1(2) of the Radio Spectrum Decision, No 676/2002/EC, 7 March 2002.

[13] Recital 22 of the Authorisation Directive 2002/20/EC, Brussels.

[14] Page 11 in P. Milgrom, *Putting Auction Theory to Work*, Cambridge University Press, 2004.

[15] "Spectrum Auction Design in New Zealand", Ministry of Economic Development, November 2005.

[16] Table C "CURRENT POPLULATION REPORTS Population Projections of the United States by Age, Sex, Race, and Hispanic Origin: 1995 to 2050", US Department of Commerce, Economics and Statistics Administration, BUREAU OF THE CENSUS, February 1996.

[17] L. M. Ausubel, "An Efficient Ascending-Bid Auction for Multiple Objects", *American Economic Review*, **94**(4), 1452–1475, 2004.

6 Spectrum trading: secondary markets

6.1 Introduction

Following the initial assignment of spectrum rights and obligations to users, whether by auction or other means, circumstances may change causing initial licence holders to want to trade their rights and obligations with others. Today this is not possible in many countries. However, in a few countries secondary trading – the trading of spectrum rights after the primary assignment – is possible. The possibility to trade radio spectrum is argued by many commentators to be a critical factor in the promotion of more efficient radio spectrum use. Furthermore, it is increasingly recognised that the flexibility afforded by trading is helpful for innovation and competitiveness.

Spectrum trading is a powerful way of allowing market forces to manage the assignment of radio spectrum rights and associated obligations and it is a significant step towards a market-based spectrum management regime. The trading of radio spectrum rights has been discussed as a policy option for many years and dates back to the seminal contribution of Coase [1]. It is widely accepted by economists and increasingly by spectrum policy makers that appropriately supervised market forces can be superior to the widely used but more inflexible command-and-control methods.

Trading of spectrum is made much more powerful when it is combined with policies aimed at promoting liberalisation[1] in use; that is allowing users to choose the use to which a frequency band is put – subject

[1] Spectrum liberalisation refers to a relaxation of the conditions attached to a spectrum licence dealing principally with services and technologies. In this chapter we focus on spectrum trading. See Chapters 7 and 8 for a discussion on the liberalisation of spectrum.

perhaps to some constraints regarding the interference that can be caused (see Chapter 7). As liberalisation provides greater flexibility, it means that spectrum trades are able to seize the opportunity for greater gains.

The FCC is taking significant steps to remove regulatory barriers and facilitate the development of secondary markets in spectrum usage rights. In 2003, it adopted its first Report and Order and Further Notice of Proposed Rulemaking (FCC 03–113), in which it established new policies and procedures to facilitate broader access to valuable spectrum resources through the use of spectrum leasing arrangements. It also streamlined procedures for approving licence assignments and transfers of control. In 2004, the FCC adopted the Second Report and Order, Order on Reconsideration, and Second Further Notice (FCC 04–167), in which it provided for immediate processing of certain qualifying spectrum leasing and licence assignment and transfer transactions. This has permitted an active market in spectrum for commercial radio services (CMRS), but the quantity of spectrum involved is small. The FCC also established the new regulatory concept, termed the "private commons", to provide additional access to spectrum in licensed bands (see Chapter 13 for a discussion on the private commons model).

In Europe the European Commission is taking the lead in promoting harmonised trading for radio spectrum where its use has a European dimension. Emphasis is being placed on certain bands below 3 GHz, where it is estimated that the net benefits from trade may be substantial. In a study for the European Commission, it has been estimated that the net annual benefits to the European Union of liberalisation and spectrum trading could be as high as €9 billion. However, the benefits of trading alone amount to only 10% of this sum.

In the UK, which has been a pioneer of spectrum trading in Europe, the regulator Ofcom estimated that the benefits of introducing spectrum trading will substantially exceed costs with net economic benefits in the range £67 million–£144 million.

In this chapter we start in Sections 6.2 and 6.3 by looking at the theory behind spectrum trading and show why it allows gains to be exploited and hence for society to improve upon its use of radio spectrum. In Section 6.4 we highlight some of the objections sometimes levelled against spectrum trading. In Section 6.5 we discuss the implementation of spectrum trading in the UK, which has taken bold measures in recent years to facilitate spectrum trading and spectrum liberalisation. In Section 6.6 we look at trading of spectrum in some other countries and Section 6.7 concludes.

6.2 Radio spectrum and market forces

Market forces work by sending price signals to market participants. For example, when demand for a good exceeds its supply, price tends to increase. The effect of this causes some buyers to lower demand or even exit the market and some suppliers to expand production and possibly for new producers to enter a market. The end result of this process is a tendency for demand and supply to be brought into equality as demand shrinks and supply expands. Could market forces work this way for radio spectrum?

Production of radio spectrum cannot be expanded – radio spectrum is given and generally fixed in quantity. This raises the question of how price signals can stimulate a supply-side response. Furthermore, radio spectrum is commonly accessible which makes it difficult to prevent use of the resource by those unwilling to pay a market price. These properties of radio spectrum, its fixed supply and common accessibility, may appear to challenge the effective operation of market forces in assigning radio spectrum rights.

The common accessibility and fixed supply of radio spectrum makes it little different from land. While radio spectrum has largely been managed via command-and-control means for decades, land has been traded on markets for many centuries. The trading of land has been enabled by the successful enforcement of property rights (largely private property rights) and by systems of land registration to document ownership rights. The trading of spectrum rights does necessitate the identification and enforcement of spectrum property rights, see Chapters 7 and 8.

In this chapter we assess both the theory and practice of spectrum trading via market forces and how such trading can work effectively in the management of the assignment of radio spectrum rights and associated obligations.

6.3 Spectrum trading, markets and efficiency

Proponents of spectrum trading argue that the design of markets intended to facilitate transactions should result in efficient outcomes. In a loose sense efficiency is achieved when the owners and/or users of spectrum are those with the highest valuations for the spectrum. Economists use efficiency in a more precise way, and three related concepts are often used: (i) Pareto efficiency (which has three components: allocative, productive and dynamic efficiency) – where resources are allocated across consumers and firms so that no firm or consumer can be made better off without making some other body worse off; (ii) Informational efficiency (where prices accurately reflect underlying value – usually of concern in financial markets); and (iii) Operational efficiency (where markets work efficiently from an institutional perspective – the issue of market design). A market is "fully efficient" when all three efficiency criteria are satisfied. For a market to be fully efficient (Pareto, Informational and Operational), a number of criteria need to be satisfied:

- thick markets – there are many potential traders,
- no market power – no potential trader is large relative to the overall market and able to affect unilaterally the operations of the market,
- well-defined property rights – radio spectrum property rights need to be clearly and transparently defined,
- full information – information about radio spectrum property rights is readily available,
- no unforeseen externalities – the use of radio spectrum by one party does not impose unforeseen costs on another party.

The efficiency conditions are discussed further below in the context of spectrum trading.

6.3.1 Thick markets

If there are many buyers and sellers in a market, it is well known that this benefits efficiency – particularly Informational and Pareto efficiency. The participation of many traders in a market results in the price mechanism processing information about buyers' and sellers' valuations more effectively – that is, price signals stimulate more efficient trading. In effect this means that the volume of trade, measured by the number and frequency of trades, is important for the efficiency of markets. However, this criterion is probably less significant in radio spectrum markets than in markets where the traded items, such as financial instruments, summarise information about other factors.

However, while the presence of many buyers and sellers is good for Informational efficiency, trade among the few can deliver gains and be Pareto efficient. For example, if firm A owns 10 MHz of spectrum and values this at $100 million, and firm B values the same spectrum at $125 million, firm A could make up to $25 million by selling the spectrum to firm B. Assuming firm B is willing to pay only up to $125 million for the spectrum, firm A can be better off (as it can sell the spectrum for more than $100 million) and firm B can enjoy a value measured as the difference between $125 million and the price at which the transaction occurs.

Figure 6.1 illustrates a situation where there are two uses/users of spectrum (A and B).

The horizontal axis is used to represent the finite quantity of spectrum available, and two demand curves are shown for competing uses/users (A and B) of available spectrum. The demand curves reflect the private values of uses/users. With trading the outcome would result in an equilibrium price P^* yielding an efficient allocation shown by the intersection of the two demand curves.

6.3.2 No market power

Pareto efficiency tends to be achieved in environments where traders are small relative to the market. In effect this means that no individual

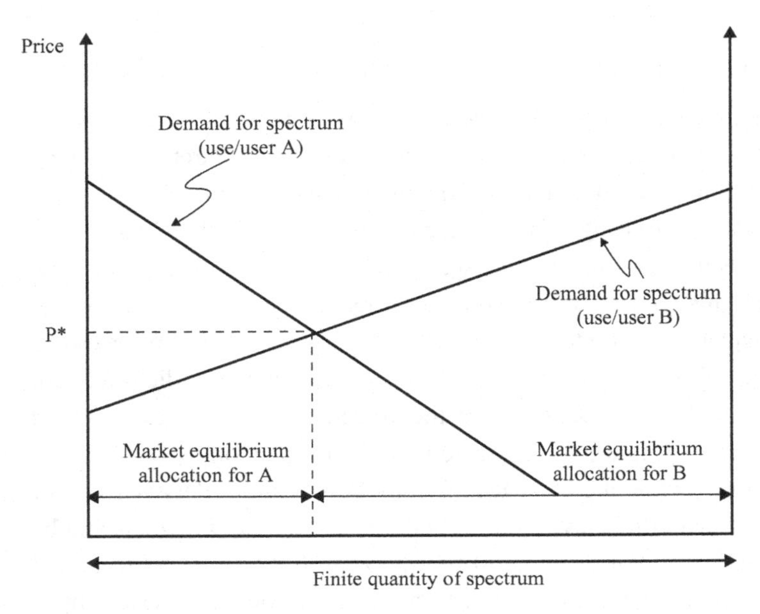

Figure 6.1. Demand curves for two users of spectrum.

trader is able to influence the market price materially – and furthermore no trader has an incentive to trade at a price different from that reigning in the market. Where a trader can "distort" a price in a market, that is influence the price at which trades occur, this can lead to a price which does not reflect underlying valuations (or more precisely opportunity costs). This is discussed further in Chapter 9.

6.3.3 Well-defined property rights

Trade in radio spectrum requires at a minimum enforceable and identifiable property rights. The reason for this can be seen by looking at the long-established markets in property in many countries. These work well partly because there tends to be a land registry scheme which details ownership records for definable parcels of land. Without a land registry scheme, trade in property would involve considerable additional costs as ownership would need to be investigated more closely.

Inevitably this would depress trading activity in the property market, as the cost of transacting would be much higher.

Spectrum markets face particular problems associated with interference. One operator's "property" in spectrum can be made valueless if another operation interferes with it, just as grazing is valueless if the owner cannot exclude other farmers' animals. These issues are analysed in detail in Chapters 7 and 8.

The operation of registry schemes helps operational efficiency by lowering the transaction costs associated with trade. From a policy perspective the costs of operating registry schemes have to be weighed against the benefits associated with lower transaction costs. Clearly where the benefits derived from institutional arrangements outweigh their costs, they should be promoted to assist the market.

6.3.4 Full information

Market forces are more effective when traders are better informed about the nature of products traded and about the valuations of traders in a market. While the identity of traders is not essential for efficiency to hold in many markets, the identity of traders in spectrum markets may provide a very strong signal about the use to which spectrum is to be put.

As spectrum usage can generate interference adversely affecting other users, which in some instances can be very costly, it is essential that as much information about spectrum assignments is made public prior to the conclusion of trades. This requirement is particularly important in situations where the costs due to possible interference are not known, or are known with imprecision, and where they are considerable. Facilitating transparency in spectrum markets would enable market forces to steer prices more effectively towards efficient levels. This might involve, for example, maintaining a public register providing information about the ownership of spectrum. Since public bodies utilise substantial numbers of frequencies, it is desirable that they participate as far as possible in such information disclosure.

6.4 Objections to spectrum trading

Spectrum trading causes a concern and encounters opposition in many quarters. Much of it is concerned with possible failure of the property rights regime, caused by interference from one spectrum user to another – Chapter 7 indicates why this worry is probably misplaced. But there are other objections, which are considered below.

6.4.1 Lack of harmonisation

Administrative allocation and assignment by the spectrum regulator allows spectrum to be harmonised internationally, and it is often accompanied by standardisation on a particular technology. The most obvious example of this is GSM, the 2G mobile standard. The European Union, via a Directive, has allocated a relatively small frequency band at 900 MHz for use by GSM technology throughout the 15 (then) members of the Union. This generated a critical mass of demand from users, equipment manufacturers and operators which permitted the technology to take off throughout the world.

In the United States, by contrast, a variety of standards were adopted, including GSM, and this was accompanied by – though not necessarily caused by – much slower penetration of the technology. (The regime of local licensing, geographic numbering and of payment by the receiver for mobile calls probably contributed to this slower adoption.)

Clearly, if an international body can correctly intuit the "right" frequencies (and the "right" technology standard) for provision of a service, moving immediately to this optimum by administrative method has advantages. But if harmonisation is adopted by spectrum regulators inappropriately, frequency bands can be sterilised for decades, and better options stifled.

Market methods thus offer an alternative form of harmonisation at a different level. If flexible pricing and secondary spectrum markets exist across a range of countries, an operator, supported by equipment vendors and suppliers of devices, can acquire frequencies in several markets, or separate operators can acquire a common frequency in difficult jurisdictions – thus achieving the benefits of harmonisation in terms of lower cost equipment as a result of economies of scale, and interchangeability of equipment across countries (so that, for example mobile voice or mobile broadcasting subscribers can gain access to services abroad).

Under this alternative approach to harmonisation, investors make the decisions, rather than regulators. If those investors have different views, they will back different approaches, and a competitive struggle could show up the best outcomes for consumers.

Clearly this is a much more uncertain and experimental process than its administrative alternative. There is no clear evidence of which is preferable – which in any case will depend on circumstances. Losing the benefits of harmonisation (if they exist) is thus a factor to consider in evaluating market methods. This is a particularly important consideration for larger countries, or groups of countries, where their combined user base is sufficient to generate the economies of scale normally needed to allow standards to succeed.

6.4.2 Windfall profits

Spectrum trading represents a means of improving economic efficiency in terms of how the radio spectrum is used. Nevertheless, the mechanism used does have implications with regard to the incentives for spectrum trading and the possibility of making windfall profits. If it is assumed that auctions result in an efficient assignment of spectrum, then there shall initially be no incentive for a trade. However, in the case of other assignment mechanisms that do not depend directly on the amount users are willing to pay, there may be an immediate incentive to trade which could give rise to certain problems.

Windfall profits accrue to owners of specific property rights without any effort or economic activity on their part. The basic premise holds that a distribution of scarce resources gives the recipients an opportunity to make a profit. If they do so as a result of commercial activity, associated with the roll out of a network infrastructure, then there are no grounds for censure. On the contrary, there would only ever be cause for concern if it were possible for the user to make excessive profits without taking on correspondingly higher risks, if profits could be made simply by trading, without engaging in any productive activity or if spectrum had been obtained by means of a non-market-based mechanism.

From a purely economic standpoint, windfall profits do not constitute an argument against spectrum trading. If, however, they are regarded as problematic for other reasons, there are various means of limiting such gains in the context of spectrum trading. First of all, usage rights should initially be assigned in an auction. Other options include a spectrum charge, effective market regulation, a windfall tax (much like a tax on capital gains) or a trading duty whereby the state recoups a proportion of the net gain when a trade takes place. Nor should it be forgotten that even windfall profits which accrue solely to the seller can also boost government finances. This is most immediately apparent if the seller in question is a state-run or state-funded institution which has purchased its spectrum rights via a market mechanism. If, for example, the armed forces had acquired some spectrum in a market and were to sell off surplus spectrum, the additional funds this would bring would generally mean that less funding was needed from other sources. Furthermore, the more efficient use of spectrum that is expected to result from trading also implies that the new users will make higher profits (assuming that the user is a for-profit entity). This in turn will result in increased tax revenues for the state (from income and capital gains tax and, if revenue also increases, from sales taxes). Once all factors have been taken into account, the issue of windfall profits, if suitable care is taken, therefore presents far less of a problem than is often portrayed.

6.5 The implementation of spectrum trading in the UK

Table 6.1 provides a chronology of the key steps taken to enable spectrum trading in the UK. Note that Ofcom regards spectrum trading as allowing users to “buy, sell, aggregate or unbundle spectrum holdings”.

In the Cave Review [2] it was stated that trading should be introduced in a way which minimises transactions costs, consistent with maintaining the integrity of the spectrum management regime. It was argued that

Table 6.1. *Spectrum trading in the UK*

12 July 2000	European Commission proposes a directive COM2000(393) on a common regulatory framework for electronic communications networks and services ("the framework directive"). Article 8.4 states "Member States **may** make provision for undertakings to **trade** rights to use radio spectrum with other undertakings only where such rights ... have been assigned ... by auction." *[emphasis added]*
November 2000	UK Government announces that it is setting up an Independent Review to look at spectrum management.
December 2000	UK Government publishes Communications White Paper, "Communications in the 21st century". It is stated "We will value the spectrum used by broadcasters and introduce new mechanisms to enable communications companies to trade spectrum."
March 2001	Professor Martin Cave appointed to head the Government's Independent Review.
July 2001	European Commission publishes COM(2001)380 to amend the proposed framework directive in which it relaxes conditions to be placed on spectrum management – permitting transfers (and not just trades) and allowing transfers of spectrum assigned by means other than auction.
Late Summer 2001	Radiocommunications Agency to publish consultation paper on spectrum trading, outlining in greater detail its proposals for spectrum trading.
March 2002	Government Independent Review *of Radio Spectrum Management* published in which it is stated "The review strongly advocates the earliest and widest application of spectrum trading possible".
March 2002	Framework Directive adopted in which article 9.3 states "Member States may make provision for

Table 6.1. (*cont.*)

	undertakings to transfer rights to use radio frequencies with other undertakings." Article 9.4 states "Member States shall ensure that an undertaking's intention to transfer rights to use radio frequencies is notified to the national regulatory authority responsible for spectrum assignment and that any transfer takes place in accordance with procedures laid down by the national regulatory authority and is made public. National regulatory authorities shall ensure that competition is not distorted as a result of any such transaction. Where radio frequency use has been harmonised through the application of Decision No 676/2002/EC (Radio Spectrum Decision) or other Community measures, any such transfer shall not result in change of use of that radio frequency".
July 2003	The Communications Act 2003 requires regulations to be introduced facilitating trading in spectrum (section 168 of the Act).
November 2003	Radiocommunications Agency and Ofcom jointly publish consultation on spectrum trading.
February 2004	Public consultation on secondary trading of rights to use radio spectrum launched by the Radio Spectrum Policy Group in Europe.
May 2004	Study on conditions and options in introducing secondary trading of radio spectrum in the European Community for the European Commission published.
August 2004	Ofcom publishes its statement on spectrum trading setting out a phased approach to implementation in 2004 and beyond.
November 2004	Ofcom publishes Spectrum Framework Review, which outlines a vision for spectrum management where market forces play an increasingly

Table 6.1. (*cont.*)

	important role in determining how spectrum is used.
December 2004	The Wireless Telegraphy (Spectrum Trading) Regulations 2004 create the legal framework for spectrum trading in the UK.
June 2005	Spectrum Framework Review statement outlines further details about spectrum trading.
During 2006	A number of trades take place. Scope of trading extended to other areas.
June 2007	Consultation issued on spectrum usage rights – a mechanism to add flexibility to trading.

licensees should have the freedom to divide and partition their licences by frequency and geography for subsequent sale. It was also argued that spectrum users should be able to lease access to frequencies to others. In these cases, the original licensee would share access to frequencies while retaining responsibility to the regulator for the conduct of the licence.

Ofcom has adopted a phased approach to spectrum trading, see Table 6.2. Different types of transfer are permitted, e.g. full or partial transfers. To strengthen the benefits of trading Ofcom has also actively liberalised use wherever possible.

Spectrum trading has been possible in the UK since the end of 2004. In 2004, roughly 1057 licences became tradable, with a further 37 000 licences in 2005 (totalling approximately 69% of all licences). Some licences contain a greater number of assignments than others. After 20 months of trading, only 15 transactions have been completed. Eleven of the transactions involved regional spectrum rights in the 28 GHz range for broadband fixed wireless services. A handful of other trades have been small in scale and connected with common base stations.

6.6 Trading in other countries

To date only a few countries have implemented spectrum trading: these are Australia, El Salvador, Guatemala, New Zealand and the

Table 6.2. *Phased approach to spectrum trading in the UK*

2004	2005	2006	2007	Other
Analogue PAMR	Wide area PBR	Emergency services	2G and 3G mobile	Mobile satellite
National paging	On-site PBR		Programme making	Radio broadcasting
Data networks	Digital PAMR		Aviation and maritime	Television broadcasting
National and regional PBR	Fixed systems at 10 GHz, 33 GHz and 40 GHz		Radionavigation	
Common base stations				
Fixed wireless access				
Scanning telemetry				
Fixed terrestrial links				

Source: Ofcom. (http://www.ofcom.org.uk/consult/condocs/spec_trad/statement/)

USA. Some of the relevant experiences from these countries are detailed below.[2]

6.6.1 New Zealand

New Zealand has adopted an approach that is based on a three-tier system of rights.

[2] The material in this section is drawn from the ITU [3].

- Management rights bestow the exclusive right to the management of a nationwide band of frequencies for a period of up to 20 years. Within this band, the manager can issue licences. They are not constrained as to the uses for which licences are issued.
- Licence rights are derived from spectrum licences that are issued by the management rights holder which allow licensees the right to use frequencies within their bands. Licences are specific to a particular use and defined in terms of transmitter sites. The management rights holder can issue licences to itself.
- In blocks of spectrum where management rights have not been created, the legacy regime of non-tradable apparatus licences continues.

The Government favoured a progressive conversion of licences to a spectrum rights regime. As the initial owner of all management rights, the Government has used auctions to make primary assignments of tradable management rights. There were 91 management rights as at February 2004, with the New Zealand Government retaining ownership of 15 of these rights, predominantly over spectrum used to provide public services.

It is left to the ensuing management rights holders whether or not to trade their rights. There are no restrictions on the activities of the operators, the number of entrants into the markets or specialised licensing requirements.

6.6.2 United States of America

In May 2003, the Federal Communications Commission (FCC) adopted a "landmark" order on spectrum leasing that authorised most wireless radio licensees with exclusive rights to their assigned spectrum to enter into spectrum leasing arrangements. Under the leasing rules adopted, licensees in certain services are allowed to lease some or all of their spectrum usage rights to third parties for any amount of spectrum and in any geographic area encompassed by the licence, and for any time within the term of the licence.

The order also creates two different mechanisms for spectrum leasing depending on the scope and responsibilities to be assumed by the lessee.

- The first leasing option – "spectrum manager" leasing – enables parties to enter into spectrum leasing arrangements without obtaining prior FCC approval so long as the licensee retains both de jure control of the licence and de facto control over the leased spectrum. The licensee must maintain an oversight role to ensure lessee compliance with the Communications Act and all spectrum related FCC rules. In enforcing the rules, the FCC will look primarily at the licensee on compliance issues but lessees are potentially accountable as well.
- The second option – de facto transfer leasing – permits parties to enter into leasing arrangements, with prior approval of the FCC, whereby the licensee retains de jure control of the licence while de facto control is transferred to the lessee for the term of the lease. Lessees are directly and primarily responsible for ensuring compliance with all FCC rules. For enforcement purposes the FCC will look primarily to the lessee for compliance, and lessees will be subject to enforcement action as appropriate. Licensees will be responsible for lessee compliance in so far as they have constructive knowledge of the lessee's failure to comply or violation.

6.6.3 Australia

The trading of radio spectrum licences was introduced in Australia in the late 1990s and was accompanied by some liberalisation. According to the Australian Communications Authority [4], 246 licences were traded between 1998 and 2004, with over 200 of these in bands below 3 GHz.

6.6.4 Guatemala

Spectrum rights in Guatemala are granted in fully transferable and fragmentable frequency usage titles (Titulos de Uso de Frecuencias or "TUF"s), which have technical limitations to protect against interference but which have no service limitations. Under the system, all spectrum that is not assigned can be requested. Following a request, the regulatory administration determines whether the request would infringe upon any other person's rights and if it does not, it opens up a period where other parties may object to the granting of the right, which must be based on a violation of the protesting party's existing right, and where other parties may seek a portion of that requested spectrum. In the latter case, the administration is obliged to start an auction. In cases where fragmentation would promote competition, the law requests from the administration that it auctions the requested spectrum in a fragmented fashion.

6.6.5 European Union

Despite fairly widespread recognition that the current regime of spectrum management operating in most of the European Union is insufficiently flexible to achieve the Union's objectives in promoting competitiveness and innovation, thus far the pace of reform is slow, although some necessary steps have been put in place, and the European Commission is promoting liberalisation across the EU.

Since 2002 the current regulatory position in respect of spectrum trading has been clear: Member States are permitted, but not required, to introduce spectrum trading. Their actions are governed by the 2002 Spectrum Decision and the Framework Directive [5].

According to Article 1(1) of the Radio Spectrum Decision the aim of the Decision is to "establish a policy and legal framework in the community in order to ensure the coordination of policy approaches and, where appropriate, harmonised conditions with regard to availability and efficient use of the radio spectrum . . . ". The Framework Directive, which governs the regulatory regime overall, states (see Article 9(3)) that "Member States *may* make provision for undertakings to transfer rights to use radio frequencies with other undertakings" (emphasis added) and Article 9(4) states "Member States shall ensure that an undertaking's intention to transfer rights to use radio frequencies is notified to the national regulatory authority responsible for spectrum assignment and that any transfer takes place in accordance with procedures laid down by the national regulatory authority and is made public. National regulatory authorities shall ensure that competition is not distorted as a result of any such transaction. Where radio frequency use has been harmonised through the application of Decision No 676/2002/EC (Radio Spectrum Decision) or other Community measures, any such transfer shall not result in change of use of that radio frequency."

Any fresh decisions made under the Radio Spectrum Decision must also be limited to "technical implementing measures". Although there is no binding interpretation of the Radio Spectrum Decision that defines the permissible parameters of a technical implementing measure, it is clear from the past practice that it can be used to adopt decisions on a band-by-band basis to harmonise the way in which a certain band is exploited – both practically and technologically – across Europe. Article 4 of the Radio Spectrum Decision could be used to achieve freedom of use on stipulated frequency bands.

In September 2005 the Commission published a Communication on a market-based approach to spectrum management in the European Union [6] which noted that a fragmented approach to spectrum reform would make it more difficult to achieve the Union's objectives. Accordingly it proposed the coordinated removal of restrictions on spectrum use in all Member States in order to promote an open and competitive digital economy.

In practice it was suggested that substantial amounts of spectrum, including roughly one-third of the spectrum below 3 GHz (the spectrum best suited for terrestrial communications) could possibly be made subject to tradable and flexible use by 2010. Clearly the Communication is a key document in which the Commission has nailed its colours to the liberalisation mast. If the plan were realised it would represent a significant step towards the desired end state set out above, even though much non-communication related spectrum, which makes up much of the remaining two-thirds of spectrum below 3 GHz, would not be covered.

In 2006 the Commission's proposed greater flexibility in spectrum management could be introduced by strengthening the use of general authorisations whenever possible [7]. Also, selected bands agreed at EU level via a committee procedure would become available for use under general authorisations, or subject to secondary trading across the EU. Common authorisation conditions for the use of the radio spectrum would also be enacted with this procedure in appropriate cases.

Unfortunately, this does not amount to a complete reform of spectrum regulation of the EU, in the direction of pan-European markets. That would require new legislation, for which there may not yet be the appetite in Europe.

6.7 Conclusion

If spectrum trading is an effective way to achieve efficient assignment of radio spectrum rights, should all radio spectrum be made tradable? In short the answer is no. Radio spectrum is not homogeneous – different frequencies can support different types of applications – and this means that there are potentially many different markets for spectrum, depending on frequencies, usage and users.[3] Regulatory enforcement of property rights may be required on a greater scale in some markets than others. Where the costs of regulation would outweigh the benefits, there may be more effective ways of managing the radio spectrum. For example, for some uses and frequency ranges it might be more effective to make radio spectrum a public good (i.e. licensed exempt spectrum).

In this chapter our focus is on frequency bands where regulatory oversight of property rights is likely to be worthwhile. Frequency bands supporting high value applications where it is difficult for regulators to determine administratively the correct assignments are ideally suited for trading. In practice, trading will confer greater benefits in circumstances where innovation is rapid and demand for final services is

[3] In many markets there are often many sub-markets. For example, the market for oil has many different grades of oil traded in different locations around the world, and there are many thousands of markets for financial instruments.

variable. Although trading of spectrum has occurred on a limited scale to date, this should change in the future as markets evolve and mature. But most significantly, the trading of spectrum will increase once restrictions on use are lifted.

References

[1] R. H. Coase, "The Federal Communications Commission", *Journal of Law and Economics*, **2**, 1–40, 1959.
[2] M. Cave, "Review of Radio Spectrum Management", Department of Trade and Industry and HM Treasury, UK, 2002.
[3] ITU, "Spectrum Management For A Converging World", 2004, available at http://www.itu.int/osg/spu/ni/spectrum/RSMBG.doc
[4] Australian Communications Authority, "From DC to Daylight – Accounting for Use of the Spectrum in Australia", Melbourne, September 2004.
[5] European Commission, Decision No. 676/2002/EC of 7 March 2002 "On a regulatory framework for radio spectrum policy in the European Community" and Directive 2002/21/EC of the European Parliament and of the Council of 7 March 2002 "On a common regulatory framework for electronic communications networks and services" (Framework Directive).
[6] European Commission, "A market-based approach to spectrum management in the European Union", COM(2005)400, 14 September 2005.
[7] European Commission, "Communication on the Review of the EU Regulatory Framework from electronic communications networks and services", COM(2006)334, 29 June 2006.

7 Technical issues with property rights

7.1 Introduction

As discussed in the introductory chapter, the key reason for managing spectrum is to avoid interference between different users. To do this, users are given licences which set out in some form their "rights" to transmit or receive. These licences can be stated in many different forms, for example:

- through a particular technology (e.g. GSM),
- through a particular use (e.g. mobile),
- through a particular set of emission characteristics known as a mask (e.g. 50 dBm in band falling by 10 dB/MHz out of band).

In existing command-and-control methods user licences are typically specified in terms of the equipment or technology they are able to use, which by careful control of neighbouring uses avoids interference. However, this approach generally does not provide users with the flexibility to subsequently change their usage as circumstances change. For example, if a licence is stated in terms of a technology such as GSM this does not allow a licence holder to subsequently upgrade their network to 3G.

As has been discussed in Chapters 4 and 6, the use of market forces could significantly enhance the value derived from radio spectrum, and one of the key ways to enable these forces is through a trading regime that also allows change of use. Achieving this requires the users' rights, often termed "property rights" by analogy with land, to be defined in a different manner. Defining these property rights in the world of spectrum is complicated because of the many different uses of spectrum and the wide range of different technologies adopted. Because of the complexity, it will not be possible in the space of this book to address all the issues and determine the optimal solution. Indeed, this will be on-going work for the regulators over the coming years. However, this chapter does explain the key issues, sets out a possible framework for property rights and shows where further work is needed.

This chapter covers:

- the key aspects of property rights,
- how easements (such as UWB) fit within a property framework,
- how interference could be managed and measurement within a framework of property rights.

7.2 Key elements of property rights

In this section we discuss the outline of property rights, without introducing the many complexities associated with interference and measurement. We then address the complexities in Section 7.3.

When considering possible forms for property rights it is worth remembering that the reason for rights is to protect neighbours from interference. In this case, neighbours can be both geographical and in frequency terms. An approach such as defining the technology that can be used does this indirectly. The specification for the technology sets out the signals that a given transmitter can emit, and the likely usage defines the density with which transmitters are likely to be deployed. The combination of these two gives the interference that neighbours can expect. However, changing either the technology or the usage might change the interference and hence cannot normally be allowed.

A different, and likely superior, approach is to define the interference directly. In such a system property rights are defined in terms of the interference that can be caused to neighbours. Any change of use or technology is allowed as long as it does not increase these levels of interference. This would appear a better approach because it directly controls interference, which is the sole reason for restricting a user's rights. It is also inherently neutral of technology and usage. Finally, by considering the property rights of its neighbours, a licence holder is able to deduce the level of interference that it might suffer and design its network accordingly. This is not something they are able to do under the current approaches to property rights. In this section we describe how such rights might be formulated.

Interference comes from two possible sources.

- Illegal transmissions in the same frequency band. These are dealt with using public and private enforcement processes and are not considered further.
- Legal transmissions in nearby bands or geographies, which spill over into the licence holder's band.

In addition, the licence holder's equipment may be sensitive to strong transmissions in neighbouring bands which it is unable to perfectly filter out. This is not strictly interference since, in theory by using a better filter, it could in principle be removed. However, in practice, there are limits on the filtering that can be adopted and such signals can degrade reception in just the same manner as in-band interference. Hence it is appropriate to consider them here as well.

Each of the key forms of interference is now considered in turn.

7.2.1 Out-of-band power

Interference occurs in all radio systems. When a transmission is made on a specific frequency, the energy transmitted normally extends across a much broader band. This is shown in Figure 7.1 for GSM where the assigned band extends by 100 kHz each side of the zero point, but emissions continue well beyond this.

The degree to which energy is radiated in neighbouring bands[1] can be controlled by:

- reducing the system capacity by lowering the transmitted data rate, which narrows the overall pattern of radiated energy,
- increasing the system costs by employing tighter filtering.

[1] As Figure 7.1 shows, interference can be received from neighbouring bands, next neighbours, and so on across a wide range of frequencies. Here the term "neighbouring" is used for simplicity.

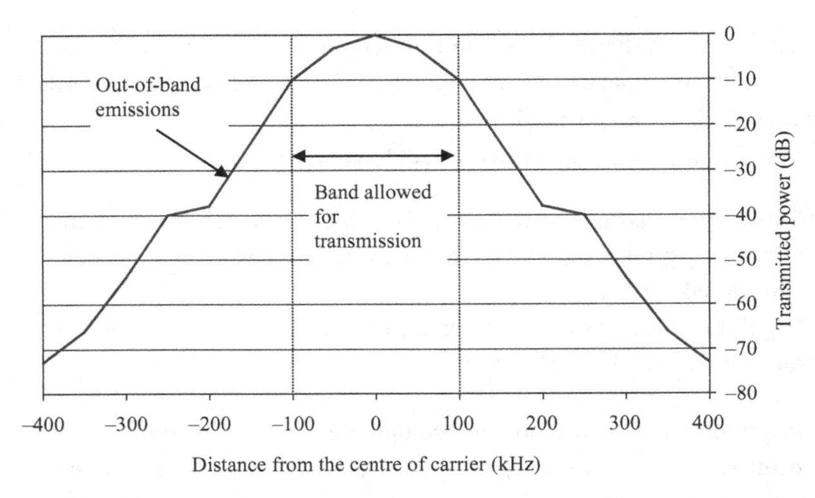

Figure 7.1. Typical radiation of energy by a cellular transmitter showing overlap into adjacent channels.

Both of these factors reduce the economic value of the spectrum to the licence holder. The overall value of a band comprising many licences can be maximised if licence holders are allowed to transmit a small but significant level of interference into neighbouring bands. Value is maximised if the value of the increase in capacity and/or reduction in equipment cost enabled by the less restrictive interference limits is greater than the value of the loss of capacity that the neighbour experiences due to the interference suffered or increase in cost of its equipment. Finding the point of maximum value is complex and is specific to technologies and uses, a point returned to later.

The amount of interference that a licence holder can emit into neighbouring bands is termed the out-of-band interference limit.[2] Under current spectrum management regimes it is normally expressed in terms of power level at the antenna of the transmitting equipment whereas under our proposed approach it would be set in terms of interference that can be caused to neighbours. The out-of-band interference limit

[2] In some cases the term "adjacent channel interference" is used.

effectively sets the interference that a neighbour can expect to suffer in its own band. It is a key component of any property right.

7.2.2 In-band power

The second key parameter is less obvious. In the same manner that it is impractical to filter transmissions very tightly, it is also impractical to filter received signals very tightly. Therefore, a radio receiver, tuned to a particular channel, will also receive signals transmitted in neighbouring channels. The neighbouring signals will be much reduced in strength by the filter in the receiver, but will not be totally removed. The ability of the receiver to filter such signals is known as the adjacent channel selectivity. Problems occur in practice when a receiver is far from its associated transmitter but close to a transmitter operating in a neighbouring band. The wanted signal will be very weak, while the signal in the neighbouring band will be very strong. Although the neighbouring signal will be attenuated by the filter in the receiver it may still be sufficiently strong to cause interference.

The current solution to this problem is to limit the power that a transmitter is allowed to emit in its own frequency band, termed the in-band power. Knowing this limit, the designer of a radio receiver can determine the maximum difference in signal levels that the receiver is likely to experience and can design the filter appropriately. Value is maximised when the combined costs of

- higher specification filters in the receivers of operator A, and
- additional base stations to achieve the same coverage as one higher power base station for operator B

are minimised. Again, this is a complex calculation that varies according to technology and usage, and even according to the number of subscribers.

The combined effects of in-band and out-of-band interference are shown in Figure 7.2.

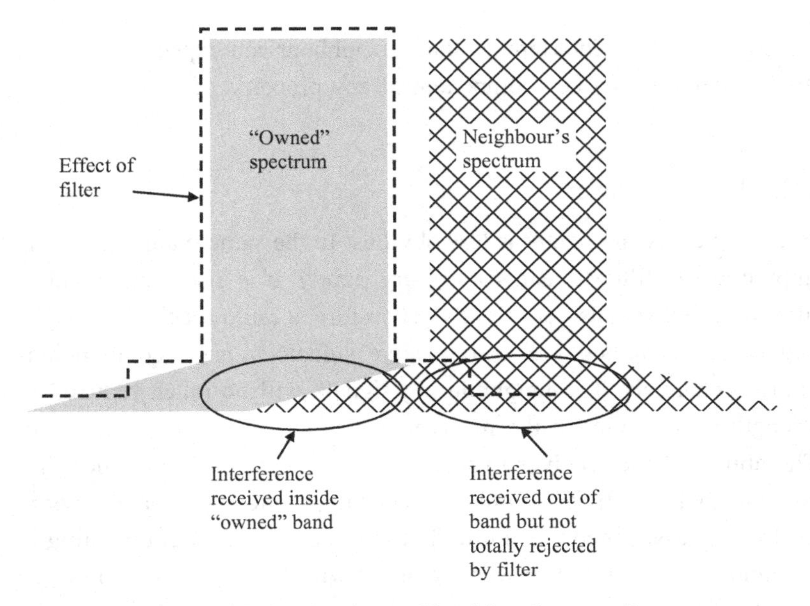

Figure 7.2. Combined in-band and out-of-band interference effects.

7.3 The problem of deployment density

As was mentioned above, the actual interference suffered by a licence holder will depend on the power emitted by the neighbouring base station and the distance from that base station. A network of a relatively small number of high power sites, such as that used by broadcast systems, would result in relatively few areas where interference might be experienced, but these areas would be relatively large. Conversely, a network of many medium or low powered sites, such as used by a cellular system, would result in many more areas where interference might occur, but these areas would be smaller.

There are a number of options to solving this problem.

- Simply set the maximum transmitter power under the assumption that the density of base stations is unlikely to change significantly. However, this may be an invalid assumption and adds risk to the deployment of networks by neighbours.

- Require every transmitter deployment to be agreed with neighbours. This would ensure that the neighbours controlled any change in interference but would likely be restrictive, bureaucratic and difficult to work in practice.[3]
- Set the interference limits on the basis of the number and size of the areas where certain levels of interference could be caused. While this would not specify where those areas actually were, it would allow probabilistic design assumptions to be made.

None of these is perfect, but our preference is for the latter. In this case an operator would be told that the interference it generates must not exceed a certain limit stated in terms of dBm/unit bandwidth for more than a certain percentage of time in more than a certain percentage of locations. In order to verify whether this was the case, a sufficiently large number of measurements would then be made across a defined unit area such as 5 km^2, for a long enough time to capture variations. In practice, modelling might be used rather than actual measurement to save time and cost.

Such an approach can be used both for out-of-band and for in-band emissions, since the interference levels experienced by a neighbouring receiver are determined in an identical manner for both.

However, the use of such a set of rights leads to a different set of restrictions on the licence holder than the existing approach of limiting maximum transmitter power. Under the current system a licence holder typically has restrictions on what they can use the spectrum for, such as fixed, mobile or a particular technology; but within those restrictions and up to a certain maximum power limit, they can deploy as many base stations as they wish. If rights were restated in these new terms then there would be no restrictions on usage or technology, however, deploying a more dense base station network might only be allowed if

[3] However, there are situations where this does occur. For example, in California the Wireless Broadband Access Network Coordination (BANC) is a group of fixed-wireless operators working together to minimise interference and maximise reliability and spectrum efficiency in the licensed exempt 5.4 GHz frequency band. See http://www.wbanc.com/

the average transmit power from each base station was reduced (which is often the case in practice anyway).

7.4 Calculating noise floor levels

Once a licence holder's transmit rights are set in terms of the interference it can cause to other users, both in-band and out-of-band, it becomes simple for a licence holder to determine the interference levels that they can expect to see within the spectrum that they own. They can do this by simply adding up the interference that their neighbours are allowed to generate in the licence holder's band. They need to add onto this any other interference such as from man-made noise or UWB. This will then provide them with a noise floor to which they can plan their system design. Hence, by implication, a system of transmitter rights, stated in terms of interference caused, also provides receive rights for a licence holder. This is an important advantage of such a proposal.

7.5 Making a property rights system work in practice

7.5.1 Setting the initial terms

The mechanism for setting the limits in the first place needs to be defined. There are two options.

- Set all the limits in all licences to the same levels (which would probably be restrictive) and then allow licence holders to negotiate them to more appropriate levels.
- Set the limits to the status quo, where it exists (for example, based on the 3G specification and typical deployment density for the 3G band), such that operators do not immediately need to embark on a negotiation process. Where the status quo is not defined, estimate it through modelling and consultation.

The end point for both approaches will be the same. It seems more appropriate to provide limits in line with the status quo and avoid unnecessary negotiation for current licence holders. Even in this case, determining limits may not be simple. For example, the 3G specifications set maximum transmitter power but not the distribution of allowed interference, which is our preferred form for in-band and out-of-band limits. To arrive at the interference distribution requires assumptions about the likely transmit power, the likely base station density and some other factors such as base station antenna gain and height and the use of power control mechanisms. These can then be input to a propagation modelling tool which can be used to estimate the interference distribution and hence arrive at the property rights. Of these assumptions, the likely base station density is most difficult. For systems still in their early deployment stage, the final base station density may be unclear, even to the operator. In this case, it is likely appropriate for the operator and their neighbours to agree on what would be a sensible interference distribution – effectively the initial limits would be set based on negotiation among neighbouring users.

7.5.2 How a licence holder determines what network they can deploy

Conversely, for an operator, once they have a licence in property right terms, they need to work out what network they can deploy. This is not as simple as a more conventional licence where they need only to ensure they are using an approved technology. Now the operator needs to ensure that they will not exceed their interference allowance. Just in the same manner as the licence terms are set, described above; they will likely do this through modelling. By using a propagation model and inputting the parameters of their preferred technology and likely deployment density they can determine if this will likely be within the limits of their licence terms. There will be some uncertainty because models can only approximate real life, but this can be accommodated either with a margin or through network measurement after deployment. In the latter case, if interference limits are exceeded the operator can either adjust their network or can discuss a relaxation in their limits with neighbours.

7.5.3 Negotiation as a mechanism for changing rights

If a spectrum user wishes to change the technology that they are going to deploy then the in-band, out-of-band or geographical limits might become inappropriate for two neighbouring licence holders.[4] The solution to this is for these neighbours mutually to agree changed limits. Practically, it seems likely that if, for example, a licence holder wanted to increase its out-of-band emissions, causing increased interference to a neighbour, it might make some payment to the neighbour to compensate them for the loss of value, or accept a similar increase in interference from the neighbour into its bands.

The information on the modified limits needs to be available to the regulator who may be called upon to investigate alleged interference and so will need to understand rights ownership in order to correctly perform its investigations. It may also be needed by potential buyers of the spectrum who will want to understand the details of the property they might buy, although this could be divulged as part of a due diligence process.

7.5.4 Shared bands and guard bands

There are some frequency bands that are shared by different users. For example, some bands that are used for satellite uplink are also shared with fixed links. By maintaining geographical separation and ensuring that the fixed links do not radiate upwards, it is possible for pairs of users like these to coexist within the same band.

Under the property rights system, each of these shared users would have a right to use the band, with in-band and out-of-band limits, but also with geographical restrictions on their usage. These restrictions might, for example, only allow them to generate low levels of interference into certain specified geographical areas.

Shared use somewhat increases the complexity when, for example, a neighbouring user wishes to negotiate a change in its interference

[4] Because transmissions spread across many assignments, it may be necessary to consider neighbours, near-neighbours, etc. For simplicity, only a licence holder and its neighbour will be discussed here.

limits. However, it does not appear to cause any significant problems for the property rights system in general.

Under this approach it is also possible to trade in guard bands. Guard bands are pieces of spectrum effectively owned by the Government but where it is guaranteed that no use will take place. Under the proposed plan, these could be turned into tradable spectrum but with relatively restrictive property rights. These rights would ensure that if anyone did make use of the guard bands they did not cause interference to others. If someone bought the spectrum they could negotiate with neighbours over whether increased emissions were possible. By this mechanism the guard bands can safely be put to a use where economically possible and without impacting on existing users of the spectrum.

7.6 UWB and property rights

Earlier chapters have already discussed UWB and the issues it raises for spectrum management. In this section we assess the impact it would have on a set of property rights.

According to our definition of property rights, the interference level experienced by a licence holder is based on the out-of-band levels for its neighbours. These should have been jointly negotiated to maximise the economic efficiency of the use of the spectrum. Any UWB transmissions, at however low power levels, increase the overall interference experienced.[5]

Equally, there may be some unwanted emissions in any band. These are generated by natural sources, such as cosmic radiation, and also by man-made sources such as electric motors. Manufacturers of non-communications equipment have to work within emissions limits for their devices, but although low these are measurable. Hence, spectrum licence holders have to take into account these other sources of noise that might impact the operation of their system.

[5] Interference levels can simply be added up. Therefore, if the out-of-band interference is 10 units and the UWB interference is 1 unit then the total interference power is 11 units. Note, however, that interference levels are generally expressed on a logarithmic scale.

By negotiating with licence holders, a UWB user could agree with them an increased interference level in exchange for a suitable payment. There are practical difficulties with this approach because of the large number of licence holders they would need to negotiate with and the potential problem of a hold-out. If UWB transmissions are deemed sufficiently valuable by the regulator, this problem could be overcome by adding an appropriate margin to the interference levels of all licences, ideally prior to issue and noting within the licence that easements were allowed. This is an approach advocated by Faulhaber and Farber [1]. Rights for UWB could then be provided on a controlled basis to ensure that the total transmitted power did not exceed the limits set by the regulator. The key here is for the regulator to assess whether economic value is increased by allowing UWB, which provides a valuable new application, but reduces somewhat the value experienced by the existing licence holders.[6]

7.7 Managing interference

Because a licence holder knows the interference level it can expect, determining excessive interference is a case of measuring the distribution of received interference and evaluating whether it is in excess of that which neighbours are allowed to cause. This can be done using standard measuring equipment. There may be practical difficulties associated with time-varying or location-varying interference requiring longer term monitoring.

Once a licence holder has determined that there is excessive interference, it can report this to the regulator for resolution, or alternatively, may, if they wish, deal directly with the source of the interference. This is very similar to the position today. According to experience in Australia, the approach of dealing directly with the interferer is the one adopted most frequently.

As we have described, there are two types of interference – illegal and legal. In the case of illegal interference, which cannot be settled by

[6] Studies by Ofcom suggest that the value generated by UWB is likely to be greater than the cost of any interference – see for example [2].

direct negotiation, the regulator should instigate the appropriate enforcement process. This can be accompanied by private legal action taken in the civil courts by the aggrieved party.

Legal interference is more complicated. It implies that the regulator has wrongly set the property rights for a neighbouring licence. In this case, the regulator needs to amend the licence terms appropriately.

7.8 A detailed look at the definition of property rights

The UK has studied property rights (which they term Spectrum Usage Rights) in some detail, and has derived a proposed set of licence terms in a consultation document. These are reproduced below, although note that they may be modified by Ofcom in the light of responses to their consultation.

For controlling emissions into neighbouring geographical areas the following could be used:

> the aggregate power flux density (PFD) at or beyond [definition of boundary] should not exceed a particular value termed X_1 dBW/m^2/[reference bandwidth] at any height up to H m above local terrain for more than P% of the time.

For controlling emissions outside of the licence holder's frequency band (that appear as in-band interference for a neighbour) the following could be used:

> the out-of-band (OOB) PFD at any point up to a height H m above ground level should not exceed X_2 dBW/m^2/MHz for more than Y% of the time at more than Z% of locations in any area A km^2.

For controlling emissions inside the licence holder's frequency band (that may cause interference to neighbouring users in frequency due to imperfect receiver filters) the same measure could be used:

> the in-band PFD at any point up to a height H m above ground level should not exceed X_3 dBW/m^2/MHz for more than Y% of the time at more than Z% of locations in any area A km^2.

Some possible numbers for these limits are also suggested by Ofcom, as follows.

Aggregate in-band PFD at or beyond geographical boundary should not exceed X_1 dBW/m^2/[reference bandwidth] at any height up to H m above local terrain for more than P% of the time.	X_1 = (based on sensitivity of services in neighbouring areas and any international agreements) $H = 30$ m AGL $P = 10\%$
Out-of-band PFD at any point up to a height H m above ground level should not exceed X_2 dBW/m^2/MHz for more than Y% of the time at more than Z% of locations in any area A km^2.	$H = 30$ m AGL X_2 = (based on service and standard "mask" for most likely technology also may be multiple values for different separations from band edge) $Y = 10\%$ $Z = 50\%$ $A = 3$ km^2
In-band PFD at any point up to a height H m above ground level should not exceed X_3 dBW/m^2/MHz for more than Y% of the time at more than Z% of locations in any area A km^2.	$H = 30$ m AGL X_3 = (based on service and maximum transmit power of most likely technology) $Y = 10\%$ $Z = 50\%$ $A = 3$ km^2

7.8.1 An example case

In order to illustrate some of the points raised above, consider the case where, owing to the introduction of digital TV, the broadcasters no longer need their entire allocation of spectrum at UHF. They decide to trade it to the cellular operators who are looking for some spectrum at lower frequencies in order to improve their coverage of rural areas. It is likely that the property rights will not be ideal for cellular operators as they will be set to allow a few high power transmitter sites rather than the multiple lower power sites that the cellular operator might prefer. Prior to buying such a licence, the cellular operator will wish to consult with their new neighbours, who may also be the sellers of the spectrum. We assume that the neighbour is willing to discuss changes to the property rights (otherwise the cellular operator would likely not proceed with their acquisition).

The downlink and uplink can be considered separately – if time division duplex (TDD) systems are used then the downlink and uplink are effectively within the same frequency band.

Downlink

Cellular systems in large countries are composed of thousands of base stations, whereas TV systems typically only have tens or hundreds. As a result, the transmit powers of the cellular system will typically be much lower than that for the TV system but users will likely be closer to a cellular base station than a TV station. There are two potential areas of interference – from cellular to broadcast and from broadcast to cellular.

- Interference from the cellular network to the TV system could occur when a cellular base station is close to a TV set. The worst case would be when a TV set is on the edge of the coverage of a TV transmitter and a cellular tower is erected close to their house and in a direct line to their TV antenna. In this case, both the in-band and out-of-band emissions from the cellular transmitter could be problematic. However, the property rights that the cellular operator inherits will likely require relatively low transmitter powers if there are many more cell sites than in the broadcasting case. These low power levels might be sufficiently low to prevent any significant interference, and indeed, more likely to prevent the cellular operator economically deploying a network. The cellular operator and the broadcaster might wish to study the maximum powers that the cellular operator can use without causing interference and modify their property rights accordingly. Or they might agree on deployment scenarios that minimise this likelihood.

- Interference from the TV to the cellular system could occur when a cellular user is close to a TV base station and so receives a high signal level. Because there are few broadcast base stations, and they are often somewhat remote, this will be a relatively infrequent occurrence. The cellular operator will have to accept this interference as part of the condition of buying the licence. This will somewhat reduce the value of the licence to the cellular operator who will not be able to provide coverage in the close vicinity of TV transmitters. The cellular operator might be able to negotiate with the broadcaster to reduce their transmitter power but in practice this is unlikely as it would result in a loss of coverage to a broadcaster with universal service obligation.

Uplink

The interference is from the users of the cellular system to users of the TV system and from the TV transmitter to the receiver in the cellular base station.

- Interference from the cellular system to the TV system could occur when a user of a cellular handset is near the antenna of a TV set. With most TV antennas mounted at rooftop level, this will be an infrequent occurrence. The most problematic cases will be when a user of a cellular phone is in the same room as a TV using a set-top mounted antenna. The initial property rights will likely make mobile transmission quite difficult as the probability of exceeding the interference thresholds in the licence, given a dense population of mobiles, would be high. Therefore, very low mobile transmit powers would need to be used which would likely make the system uneconomical. As a result, the cellular operator will wish to negotiate increased power limits. The in-band and out-of-band limits for the cellular phone need to be set to minimise the likelihood of this interference. However, some judgement needs to be used. If the limits were set such that there would be no noticeable interference, even were the handset placed next to the TV antenna, the emission limits might be so restrictive as to prevent the cellular system working properly. Instead, limits might be defined so that there was little noticeable interference when the handset was more than 1 m from the TV antenna. Alternatively, guard bands could be used to separate the two applications further in frequency terms.

- Interference from the TV system to the cellular receiver might be the most serious case of all. Here, signals from a TV transmitter are received by a cellular base station, which is trying to receive relatively weak signals from mobiles. Because both base stations might be mounted on hilltops, the propagation from one to the other might be good, resulting in a strong interfering signal at the cellular base station. Again for reasons of practicality and universal service obligation the cellular operator will likely need to accept the power transmitted by the broadcaster. To mitigate the problem, the designer of the cellular system might select their cell sites to be as far away as possible from known TV transmitter sites, or insert filtering that can ease the problem.

The discussion above illustrates that there may be up to four situations of interference to consider – from system A to system B and from system B to system A in both the uplink and the downlink.

Having studied these cases the cellular operator might enter into discussion with the neighbouring broadcasters, putting forward some proposals as to how the property rights could be changed. The cellular operator might in compensation offer some payment or other concession to the broadcasters. Once these negotiations had concluded the cellular operator could determine the type of system they could deploy and build a business case. This would allow them to understand what they could pay for the spectrum that they wished to purchase, and if viable move ahead with the acquisition.

7.9 Summary

In this chapter we have discussed the technical issues associated with the formation of property rights that would assist trading and allow change of use. We suggested the following.

- A set of property rights is critical to facilitate trading and enable change of use.
- The key components of property rights comprise limits on the levels of interference that can be caused in three cases: in-band, out-of-band and at geographical boundaries. From these limits, neighbours will be able to calculate their likely interference levels.
- These rights and obligations will need to be carefully defined since the interference experienced by one user depends on the rights granted to other users.
- UWB can fit within this framework of property rights; however, the regulator will need to make a decision, likely based on the economic case, as to whether it should be allowed.
- Introducing such a system will likely be complex and time-consuming.

References

[1] G. Faulhaber and D. Farber, "Spectrum Management: property rights, markets and the commons", in F. Craven and S. Wildeman (eds.), *Rethinking Rights and Regulations: Institutional Response to New Communications Technologies*, MIT Press, 2003.

[2] Ofcom's Research Papers on this topic can be found at http://www.ofcom.org.uk/research/technology/archive/cet/uwb/uwbpans/#content.

8 Economic issues with property rights

8.1 Creating property rights: economic aspects

Moving to a regime for secondary trading (as well as primary auctioning) of spectrum requires, as well as a clear technical definition of rights, a clear economic definition. As an illustration, spectrum licences in the UK have traditionally been held on an annually renewable basis, the licensee having further unspecified protection based upon a "reasonable expectation" of longer tenure. This lack of specificity would clearly create major and avoidable uncertainty in a spectrum market, and deter both transactions and the collateral investment necessary to put the spectrum to work. It is thus universally recognised that a trading regime requires a detailed specification of rights.

In principle, these rights can be embodied either in a tradable licence to use spectrum, or to install spectrum-using apparatus, or as directly owned property. In practice the tradable instrument in most jurisdictions is a transferable licence, and our discussion below is based on this approach, although we sometimes speak of "trading spectrum" rather than "trading licences".

This chapter discusses some of the issues in the definition of licence conditions (construed as above). Section 8.2 sets out some of the basic economics of property rights. Section 8.3 considers key issues in how rights should be defined from an economic or commercial point of view. Issues concerned with technical (interference-related) aspects of property rights were dealt with in Chapter 7.

8.2 Principles for the allocation of property rights

Most property, for example land and physical assets, has a clear owner, whose rights are protected by private and public law, including restrictions on theft.[1] In Europe a government's abilities to change those rights are increasingly restricted by human rights legislation. Other resources lie outside the conventional system of property ownership, typically being subject to some kind of administrative allocation by a public or quasi-public body, sometimes accompanied by an unspecified "reasonable expectation" of continued access. In certain cases, such resources are traded in a *de facto* market – landing rights at airports are an example.

When such resources are converted into tradable property through legislation, the government or regulator has a rare opportunity to define property rights. If the rights created are perpetual ones, it is a "once and for all" chance. If the rights are confined to a period, the resource may revert to the government which can then reshuffle the associated rights and reissue them to others, for example by an auction.[2]

One starting point is that in defining rights for spectrum the government or regulator should be guided by considerations of economic efficiency. The underlying logic for this criterion is that spectrum is an input, and that – whatever interventions governments want to make on the composition of final outputs, by taxing or subsidising marketed goods or by direct public provision – they should always seek to use inputs efficiently. Doing otherwise will avoidably restrict the economy's potential.

This approach does not rule out the direct reservation of spectrum for particular purposes, provided that the allocation chosen takes into account the opportunity cost of the spectrum so assigned. It may imply, however, that, especially in the case of marketed services produced by the private sector, spectrum should be allocated and assigned on a

[1] The economic property rights an individual has over a commodity (or an asset) can be defined as "the individual's ability, in expected terms, to consume the good (or the services of the asset) directly or to consume it indirectly through exchange".

[2] The assignment of 3G licences with a limited life is an example of the latter case.

market basis, in view of the market's ability to direct resources to their most profitable users in a decentralised way which exploits the private information of market participants – see Chapter 6.

From this starting point, property rights should be designed to ensure that spectrum markets lead to the most efficient use of spectrum – in the dual sense that the spectrum gravitates to the most efficient use and is deployed by the most efficient firm.

What should be done to achieve this end? For the past 40 years, economic discussion of property rights has taken place in the shadow of a deeply ambivalent general result known as the Coase theorem. This was first set out by Ronald Coase in 1960 [1], and its publication followed the same author's suggestion in the previous year that spectrum should be turned into property and bought and sold in the market place. Subsequently, the theorem has been formalised as follows.

> "Whatever the starting point, trading within the framework of a system of property rights will lead to an efficient allocation of resources provided that:
>
> - the property rights are fully and precisely defined
> - there are no transaction costs
> - efficiency is defined without reference to the distribution of income."

By way of interpretation, an efficient allocation is one where resources are used where they yield the greatest value to end users of private and public sector services. The first condition requires that there be no ambiguity about who owns what. Transactions costs, mentioned in the second condition, refer to the costs of, or impediments to, the transfer of property rights. Most obviously these are monetary, such as legal costs, but they also include instances where conflicting objectives among trading partners make it impossible to carry out a trade which improves efficiency. The third condition means that we are equally prepared to recognise as efficient a regime in which all assets are held by a few large corporations and their shareholders as one where they are more equally held. (This point is held over until Section 8.4 where we discuss windfall effects of creating spectrum property rights.)

The ambivalence of the Coase theorem resides in the tension between its bold general statement and the conditions under which it is true. An optimistic interpretation in relation to spectrum trading would therefore go as follows.

> Once spectrum owners have tradable rights, they will have a monetary incentive to move them to the most efficient use. It does not matter, then, if the government initially allocates interference rights inefficiently between two adjacent licensees, as they will simply rectify the mistake by a bargain between them, involving the buying-out of one party's rights by the other. Once rights are defined, such transactions between neighbours can be done economically. This means that the design of property rights is a non-problem – owners will quickly unpick any mistakes the government makes.

The alternative pessimistic version goes as follows.

> The principal irrelevance of the Coase theorem stems from its assuming away of transaction costs. It is of fundamental importance to start out with a "nearly optimal" set of rights because it will be difficult to change them. This is mainly because (a) many changes will affect a large number of spectrum users, and it will be very costly to get them to agree (the "large numbers" problem) and (b) where only a few spectrum users are involved, each will try to appropriate the gains from trade, so no agreement will be reached (the "small numbers" problem).

We veer more towards the second view on the costs of trade, which implies a need to try to optimise property rights before the market opens. This is anything but easy. There are basically three non-exclusive approaches.

- Calculate and establish by *fiat* an optimal configuration of rights; this might involve, for example, putting the onus to avoid interference on the so-called "least cost avoider" – i.e. giving any disputed right to the party that has the fewest alternatives to using the right in question, rather than the party which can easily work around its lack of rights.

- Where that is not possible – or simultaneously – invoke a principle designed to minimise transactions costs; here the main candidate in the case of interference is to distribute rights *away* from the party that can most economically incur the transactions costs to rectify any mistakes.
- Anticipate and manage the risks associated with failure to reach the optimum solution, by causing the right to revert periodically to the government or regulator.

These approaches can be illustrated by reference to a crucial problem in interference management; that of establishing a "noise floor" such that other spectrum users can cause interference below that threshold level, but the spectrum owner has rights against interference above that level. The discussion is clearly relevant to UWB, described in Chapters 2 and 3.

The first approach would require the regulator to establish the noise floor which would maximise efficiency, and which would emerge in a frictionless market without transactions costs. This would involve weighing up the costs and benefits of different allocations of rights, and choosing the one with the largest net surplus. This is partly based on the premise that it is practicable, where a conflict between users is involved, to identify the "least cost avoider" – the party that can change its conduct most economically to avoid the conflict – and give the right to the other party.

Suppose, as an illustration, that this is a dispute over ownership of interference rights between a TV broadcaster and a 3G operator. The TV broadcaster has little ability to change its network because of the few masts it owns on carefully selected sites and as a result of the fact that most viewers will have antennas oriented towards the nearest mast. However, a 3G operator which had already deployed a network at 2 GHz where signals travel shorter distances than at TV frequencies, could select from among many of its cell sites when deploying in the band. In this instance, the cellular operator has much more flexibility than the TV broadcaster to forego the property rights, and by the "least cost avoider" principle, if they go to the broadcaster no transaction will be necessary.

The second approach focuses not on searching for the optimum but upon facilitating "improving" transactions. On this principle, if one party can promote a trade in rights at low transactions costs, it should not be assigned the right, on the ground that it is best placed to initiate the corrective trade. For example it may be cheaper for a single spectrum user to buy out multiple users, than the other way round, as forming a coalition of multiple users to buy an asset of common value to them all will come up against the "free rider" problem – none of them will want to contribute to the joint cost.

Both of these principles may involve the choice of different solutions in different frequencies, where technical and demand conditions may pull in different directions. The regulator will therefore have to decide whether to go for a general solution or for a series of particular solutions. The latter is more difficult, but may more accurately reflect the post-trading equilibrium. The optimal solution will also vary over time, which presents a problem: should the calculation be done on the basis of today's or tomorrow's technologies?

The third principle is more of a fallback or an insurance policy. Suppose transactions costs lock spectrum up in demonstrably inefficient uses – for example, bands are rendered useless for new technologies by a single hold-out owner. How can this be avoided? The one answer is by abridging property rights in favour of a form of "compulsory purchase" to unlock the gains. However, any such power would have to be circumscribed, as the prospect of compulsory purchase, even with compensation, would chill incentives to invest in spectrum and in collateral assets. Another more comprehensive approach would be to set a limit on the licence period, again with probable disincentive effects on investment, especially as the hand-back date approaches.

8.3 Underlays and overlays

One of the most controversial issues in spectrum management concerns underlays and overlays. Underlays are exemplified by UWB, which operates under the noise floor of other services. An overlay connotes access by unlicensed users to a licensee's spectrum, provided that no interference is caused to the licensee. Such access is by analogy with what is sometimes called in property law called an easement. We deal with underlays first.

In principle, UWB could be utilised in at least three ways. First, one or more separate geographical licences could be carved out beneath any existing noise floor and assigned on an exclusive basis. Or the same space could be carved out, and made licence-exempt. Or an obligation could be imposed on any prospective user of UWB to negotiate an arrangement with all licensees under whose noise floor it proposed to operate.

The last option would almost certainly fail because of the transactions cost incurred in negotiating with countless licensees. Choice between the licensed and licence-exempt modes of permitting UWB should hinge upon a calculation of the risks of interference, either between UWB users or from UWB users to other licensees. It is necessary here to conduct a risk management exercise to establish what would be the consequences in the future if the multiplication of UWB users ultimately led to either of the interference problems noted above. The conclusions reached so far by regulatory agencies (notably the FCC and Ofcom) favour creation of a traditional commons with very strong power limitations.

Finally, there is the question of overlays, or access by users to spectrum licensed to others. In principle, this could be made generally available. Indeed, the European Commission's recent proposals on spectrum reform [2] seem on some readings to contemplate such a general right of access, when they say that:

"A new system for spectrum management is needed that permits different models of spectrum licensing (the traditional administrative, unlicensed and new marked-based approaches) to coexist so as to promote economic and technical efficiency in the use of this valuable resource. Based on common EU rules, greater flexibility in spectrum management could be introduced by strengthening the use of general authorisations whenever possible" (p. 7).

This clearly raises fundamental issues of spectrum management and the design of property rights. Within one given frequency, transaction costs would not preclude bargaining between an original licensee and potential secondary users. There is also a concern that unlicensed entrants, lacking security of access, would not be in a position to make collateral investments, or to offer adequate assurances to their customers of continuity of supply and quality of service. Alternatively, they might establish de facto some squatters' rights of a contestable nature, which would prevent the licensee from being able to exploit its asset to the full.

These questions have been widely debated in recent years. Notably, Faulhaber and Farber [3] have argued in favour of unlicensed non-interfering overlays or easements of this kind, arguing – *inter alia* – that absent imposition of such a regime, licensees will not be prepared to supply access to secondary users by entering into appropriate contracts with what might be their rivals in downstream markets.

Baumol and Robyn [4] conclude the opposite, basing their argument in part upon the analogy of a successful market for licensing intellectual property. They assert that:

> "the argument favouring market accommodation of such emissions rests on the now familiar position that spectrum rights holders have an incentive to act in ways that result in (approximately) optimal use of spectrum space. There is reason to question whether a governmental rule without price and profit incentives will be able to match the performance of a market regime" (pp. 61–2).

Absent any experience of the non-interfering easements regime, it is hard to discriminate on a priori grounds between the two approaches. The one economises on transactions costs, while the other eschews use of the price mechanism to ration access. As often in economics, it is unlikely that there will be a single solution, with the same regime optimal in all frequencies. Ideally, the choice of regime would be determined by some kind of quasi-market testing of a secondary commons. But the design of such a mechanism is very challenging. The best way forward may be to undertake some limited testing of the non-interfering easement approach in likely looking frequencies. This could provide a better empirical base for an evidence-based decision.

8.4 Defining property rights for spectrum: commercial and economic issues

We now move on to some practical problems which a regulator will encounter in defining rights.

8.4.1 Leasehold or freehold?

Spectrum licences in many countries have traditionally been finite, but indefinite. The regulator has exercised its right to give notice to licensees, sometimes offering to transfer them, at the regulator's expense, to another band. Such arrangements are part and parcel of the command-and-control model, in which decisions about the abandonment of technologies and re-use of spectrum are taken by administrative means.

In its report on spectrum management, the Australian Productivity Commission noted the coexistence in Australia of apparatus licences, fixed to a particular use and technology, and spectrum licences, with interference limits and no other restrictions on technology and use. The former are time-limited, and require arrangements for the government to resume the spectrum. But there is no such need in relation to the latter. For this and other reasons, the Commission favoured conversion of apparatus licences into spectrum licences.

This still leaves open the issue of the duration of such licences. When licences in the UK have been assigned by an auction process, it has been necessary to provide security of tenure to the licensees, which in some cases have paid many billions of pounds for their rights. On one hand, since the licences do not permit change of use, some termination date is appropriate. On the other hand, too short a period will discourage investment in collateral assets and too big a proportion of the licence period will be blighted by the licensee's need to run down its assets. The choice of a period of about 20 years for auctioned licences suggests a belief that a lengthy period may be required to create optimal investment incentives.

A flexible spectrum licence is not, however, subject to restrictions on change of use: as a result, there is no natural upper limit for its duration. So why should licences not be perpetual?

The holder of any perpetual rights would, of course, be entitled to lease all or part of those rights to any other person for a temporary period if it were profitable to do so. Thus there would be flexibility but no automatic temporal restriction. Administrative costs would be diminished, as there would be no need for the government to take back spectrum and re-assign it. As compared with a 20-year licence, the value of a perpetual licence would not be much greater, so the revenue-raising motive is not a powerful one.

Perpetual licences also encourage firms to undertake the long process of developing bands, which may involve the development of new technologies. This is an example of an activity which might, pejoratively, be called speculation (see Chapter 9). But provided the operator has no market power in spectrum and cannot raise its price by restricting supply, the effect of such conduct is generally beneficial.

Temporary licences can cause some problems. For example, in many countries in Europe the 2G licences expire over the next decade. It is far from clear what to do when this happens. Reclaiming the licences could conceivably leave the country without 2G phone service. Both operators and regulators will spend much time and effort considering and lobbying on this issue in the coming years. An indefinite flexible licence would avoid all this effort.

On balance, we support the notion that spectrum licences should normally be perpetual. Making them so will not only enhance the tradability of spectrum, but also assist in ensuring that it is used efficiently. Governments are naturally cautious about making irrevocable decisions about public assets, but in this case the arguments for doing so seem strong.

8.4.2 Can encumbered spectrum licences be traded?

In principle, there is nothing to prevent a spectrum licensee, which has leased spectrum to another firm, selling its licence to a third party, subject to the protection of the rights of the lessee. This is a natural and, in other areas, conventional form of transfer of property rights.

In a transition phase, the government may also find it desirable to auction long-lasting or perpetual rights in spectrum subject to existing licensees with time-limited rights. It would then be up to the successful bidder to negotiate with these licensees over the repurchase of their rights, or to allow the licences to run their course. No further regulatory intervention would necessarily be required.

8.4.3 Should vacant spectrum be sold?

Much spectrum in many jurisdictions, especially in the high frequencies, has not yet been licensed. Typically the regulator, acting subject to its international obligations, waits until one or more firm has expressed interest in using the spectrum for a particular purpose (which, under the terms of the licence, cannot be varied).

Under a tradable regime, however, spectrum licences can be issued to firms without restrictions on use or resale. As a result, licences could be issued for all spare spectrum, allowing the market to leave it vacant or apply it in some use over time.

The acquirer of such spectrum, whether as intermediary or a final user, would then have an incentive to undertake development work on new technologies – work which it might not undertake without guaranteed access to the spectrum. A good example of putting this into effect is Ofcom's current major programme in the UK for the release and licensing in tradable form of significant amounts of currently vacant spectrum.

8.4.4 Dealing with the hold-out problem

Does a regulator need a tool to protect against inefficient use of spectrum resulting from hold-outs? The scenario involves a new use of a significant band of spectrum. The operator acquires 99% of the requirement, but one owner – the hold-out – tries to extract a significant part (or all) of the net present value of the new project. Should the regulator have the power to acquire the spectrum compulsorily at its value in current use and pass it on to the operator which wants it?

This kind of "physical taking" has been subject to much legal and economic analysis, especially in the United States. There are cases where a public body has exercised its power to hand over the asset to a private owner, but these are exceptional. More usually the asset is used in the public interest, to acquire land to construct a bypass for example. The literature has also addressed the question of what compensation should be paid to the original owner.

Our issue is whether the regulator needs such power in relation to spectrum used either for public or private purposes. Firstly, it must be recognised that having such a power imposes costs. These include not only the legal costs of proceedings but also the impact on investors of the uncertainty created[3] and the costs of countermeasures a party subject to compulsory purchase might take, such as excessive division of the spectrum.

Secondly, to what extent are there cases where a small amount of spectrum is indispensable to the development of a new service? If the operator could work round the absence of a piece of spectrum, a hold-out strategy would not be worth it.

Thirdly, only an inexperienced negotiator would end up as a long-term hold-out, since the potential purchaser of the spectrum would be normally willing to make an offer above the value of the spectrum in the existing use. A hold-out who never makes a deal never makes a profit.

These considerations suggest that a power of compulsory purchase has certain costs but uncertain benefits. It would be necessary to rehearse some detailed scenarios, plausibly leading to major detriments, before a "taking" power could be justified. It should also be borne in mind that regulatory agencies are subject to a variety of pressures, and giving a regulator discretionary power to take a company's property and give it to another company would subject it to considerable lobbying, and might even bring its motives into question.

[3] This is sometimes called the "demoralisation cost".

This argument applies equally to the acquisition of spectrum for non-commercial public services too. If the aim of spectrum management is economic efficiency, the public user, taking proper account of the spectrum's contribution to the value of public services, should be prepared to pay the market price. The only exception we see to this principle is a national security use, where governments in any case have wide power to overrule property rights of all sorts.

8.4.5 Should spectrum licensees pay an annual charge to the state?

It would be open to the regulator when designing the rights associated with the spectrum licence to impose an obligation to pay an annual fee – whether the licence was for a period or perpetual. Is this desirable? The question arises particularly because an administrative pricing system for some reserved bands will run alongside tradable spectrum (see Chapters 11 and 12).

Clearly there will be an easily calculable relationship between the capital value of any block of spectrum and the scale of the annual payment. Market transactions will reflect the obligation. This does not mean, however, that an annual change would have no effect.

Where current short-term licences are directly translated into long-term licences, an annual charge would reduce the capital gain the licensee would receive. If the annual fee rose above the marginal value product of the spectrum, no-one would want to be the licensee, and the spectrum would be unused.

If the value of the annual fee were changed on a discretionary basis, the licensee would be subject to an unpredictable change which would introduce uncertainty and reduce the value of spectrum. This would happen, for example, if the annual charge were the same as or linked to administrative prices which changed over time to reflect changing opportunity costs.

In our view, a low "ground rent" would have no significant effect, apart from imposing collection costs. A higher rate based on an estimate of opportunity cost would produce a conflict between the need to adjust it periodically and the uncertainty created by such re-adjustment. For this reason, the best justification (in our view) of a charge is to deal with the windfall gains problem.

8.4.6 Should property rights in the transition to a market regime be limited to avoid windfall gains?

When the UK announced plans to create a spectrum market the Government indicated that it would not require existing users to undergo an auction to provide their existing services on the same frequency. When such spectrum became tradable, property rights would be assigned to existing licensees. This creates the prospect of windfall gains, which would be politically unpopular. The issue was thus raised as to whether some compensating reduction in property rights should be implemented.

The annual charge mentioned above is one means of retrieving for the state part of the value of property rights granted through the creation of tradable spectrum licences. Other approaches, such as taxes on capital gains or limiting the duration of licences, would probably have more baleful consequences as they would discourage trading.

Such measures should be tested against the criteria of minimising the impediments they create to efficient spectrum use. This generally favours "once and for all" irreversible payments which do not have an effect later on behaviour at the margin. But it would be wrong to make a once and for all charge based upon the existing configuration of spectrum use. The point of spectrum trading is that it will permit a re-allocation of spectrum which reduces existing scarcities. No firm would be willing to pay a charge to convert its current licence into a tradable licence on the basis of existing valuations. Indeed, imposing any significant charge runs the risk that it will hamper trading, as licensees will decline to acquire the new rights.

8.5 Conclusion

Our analysis has enabled us to reach a clear recommendation – a new property right should be established for spectrum trading, or more precisely the trading of spectrum licences. Our conclusions flow in part from technical consideration and in part from economic consideration, the latter relating to transaction costs and investment incentives.

- Easements should not generally usually be allowed at this stage, but rights should be assigned in ways which take account of the economic value, and interference potential, of new underlay technologies such as UWB.
- Licences should ideally be perpetual.
- Vacant spectrum should be placed in the market place (subject to international agreements).
- A compulsory purchase power for spectrum should be sharply confined, possibly to national security needs.
- Spectrum licensees should not be subject to any other restrictions that discourage efficient trading.

References

[1] R.H. Coase, "The problem of social cost", *Journal of Law and Economics*, pp. 1–44, 1960. Coase is also the author of the first major article on property rights in spectrum: "The Federal Communications Commission", *Journal of Law and Economics*, pp. 1–39, 1959.

[2] European Commission Communication, "On the Review of EU Regulatory Framework for electronic communications networks and services", COM(2006)334 Final, 29 June 2006.

[3] G. Faulhaber and D. Farber, "Spectrum Management: property rights, markets and the commons", in F. Craven and S. Wildeman (eds.), *Rethinking Rights and Regulations: Institutional Response to New Communications Technologies*, MIT Press, 2003.

[4] W. Baumol and D. Robyn, "Towards an Evolutionary Regime for Spectrum Governance: Licensing or Unrestricted Policy", AEI Brookings Joint Centre for Regulatory Studies, Washington DC.

9 Competition issues relating to spectrum

9.1 Introduction

Spectrum is an essential input into many services which are highly valuable in both a commercial and public service sense. In Europe it has been estimated that radio spectrum contributes between 2% and 3% of GDP. Public services, such as defence and the emergency services, are usually provided by a public agency and made available free to the population. There is thus no problem over monopolisation leading to excessive pricing. But commercial services are sold into a market, which makes it necessary to oversee that problems of anti-competitive behaviour do not arise.

It follows that, where spectrum is an essential input into a service market such as mobile telephony, a firm which controls the necessary spectrum also controls the downstream service sold to end users. To some extent, there may be non-spectrum-using alternatives to the wireless ones – fixed line calls instead of mobile calls and cable television instead of terrestrial or satellite broadcasting. But in many cases either these services are not good substitutes for the wireless ones, as fixed calls are not always a good substitute for mobile calls, or the relevant competing platform, such as cable TV, may simply not exist.

Figure 9.1 shows the role of spectrum in the value chain, based upon the examples of mobile communications and broadcasting. It shows, in particular, that competition in the provision of services to end users can be inserted into the value chain at other points than by licensing spectrum to several operators – for example by the intervention of mobile virtual network operators (MVNOs),[1] or by allowing a firm to "resell" a

[1] An MVNO is an operator that combines access to a mobile operator's spectrum and some of its network operations with its own assets to retail mobile telephony to its customers.

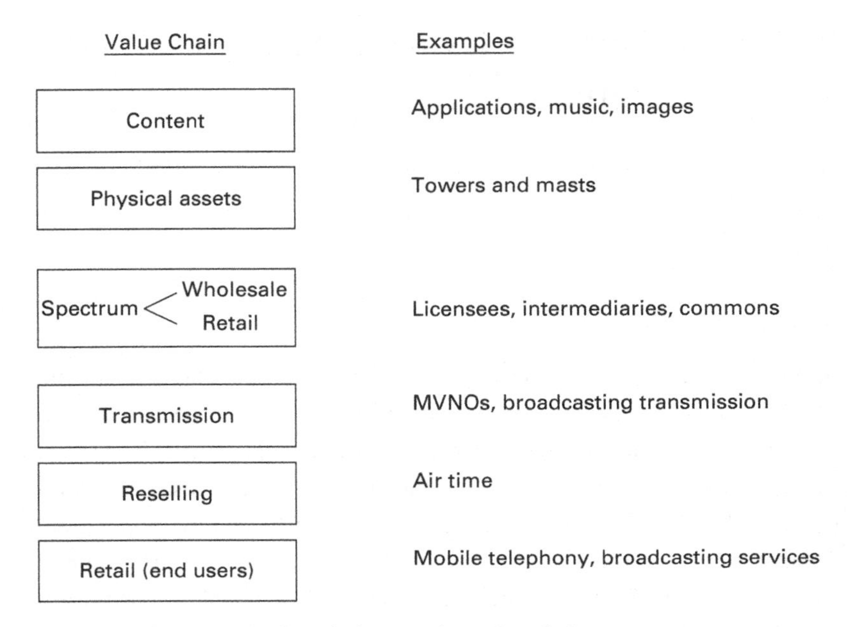

Figure 9.1. Spectrum in the wireless service value chain.

mobile operator's call minutes. But a more complete form of competition, which probably provides more benefit to end users, comes from licensing spectrum directly to operators, who can then choose how best to provide services which meet their customers' needs.

Competition in spectrum-using services generally benefits consumers by offering them lower prices and higher quality of service, as well as choice of supplier. This has been demonstrated very comprehensively in mobile telephony, where the introduction of a second, third and subsequent operator in countries as diverse as large Western European ones and small Caribbean nations has brought prices down significantly. There are, therefore, good grounds for adopting a spectrum management regime which allows competition to develop. How many competitors emerge will depend upon the size of the market, income levels and so on. But world-wide experience shows that it is much easier to generate competitive outcomes in communications markets which rely on spectrum than in those which rely on fixed networks.

Competition in services can be achieved both by command-and-control or market methods of spectrum management (or the commons model when the circumstances allow it). Section 9.2 discusses the command-and-control model; Section 9.3 the market model; and Section 9.4 the role of spectrum caps.

9.2 Competition issues in a command-and-control regime

If the spectrum regulator assigns a pre-determined number of licences to operators to provide a particular service (licences which are not tradable or flexible), it is effectively determining the maximum number of firms in the market place. It is not, however, fixing the minimum number of firms, as not all licensees will necessarily offer service. They may choose to return their licences, as happened in a few European countries where several 3G licences were returned shortly after issue in 2000–2. Where licences are assigned by an auction process (see Chapter 5) may change the choice of licensees but it does not change the regulator's influence on the market structure.

This discretion over market structure can be used by the regulator in various ways. If the goal is to maximise spectrum revenues, a goal which is unlawful in the European Union, this may best be achieved by auctioning a single licence to provide, say, advertiser-supported or pay television or mobile telephony. Consumers then pay the price by being over-charged for the relevant service. Other regulators may pursue a goal of promoting competition. This will involve them at the time of assignment of licences in (a) forming a conjecture about how many firms will be able to compete, using their own wireless networks, in the market place for services to end users, and (b) prohibiting any operator from being given more than one licence.[2]

In the UK, for example, the assignment of 3G licences in 2000 was preceded by a period of analysis as to how many competing networks

[2] This is necessary, because the most profit can be made out of the downstream services market by a monopolist. Hence the highest total bid for all licences would be made by a firm which expected to acquire all of them.

the market could support. There were then four 2G operators, and it was decided to create five 3G licences, at least one of which, therefore, would go to a newcomer. The five licences were issued and five 3G networks constructed.

In Germany, a more flexible approach to the structure of the 3G market place was made possible by enabling an auction to yield between four and six licences. The outcome to the auction was the assignment of six licences.

After the initial assignment of licences under a command-and-control spectrum management regime, the industry structure is highly constrained. Some licensees may not build out their networks and either retain their licences or be obliged to give the spectrum back under a "use it or lose it" licence condition. In the latter circumstances, the regulator can do the following.

- Create a new competitor, by using either a beauty contest or an auction. Its freedom of manoeuvre may be constrained, however, by obligations it owes to existing licensees not to lower the price for the re-issued licences, as existing licensees will not want a new competitor to get access to a cheaper licence.
- Assign returned spectrum to existing licensees; this removes the possibility of further entry at a later stage.
- Reserve the spectrum for other purposes.

The regulator may also, when auctioning a fixed number of licences, have explicitly or implicitly accepted a legal obligation not to issue any more licences for the same purpose for a specified period. The UK government entered into such a commitment at the time of its 3G spectrum auction in 2000. It is designed either to encourage network investment or to increase revenues. If the period of self-denial is long it can give market power to existing licensees, effectively providing a legal barrier to entry. As a result, consumers can be disadvantaged.

Once the assignment of licences has been made, under a command-and-control system the regulator may have powers to authorise the transfer of a licence to another owner, or to an existing operator.

Regulatory approval may be required for any merger between operators, and the return of some spectrum may also be entailed.

The outcome in terms of efficiency of spectrum use and benefits to end users depends on how the service market develops. If it fails to prosper, an excessive amount of spectrum will be tied up inefficiently. If market growth exceeds expectations, capacity may be in short supply and the number of competitors may be suboptimal. Either of these will harm end users' interests.

In summary, command-and-control gives the spectrum regulator the opportunity to shape the structure of a market by choosing the number of licences to issue. Thereafter, that structure risks becoming ossified as the spectrum regime does not provide the opportunities for firms to adjust to new information which appears on the market.

9.3 Competition issues under a market regime for spectrum management

The main characteristics of a market regime are set out in Chapter 4. Here the focus is on competition problems which may arise. As Section 9.1 has shown, a firm which controls the spectrum suitable for (or allowed to be used for) a particular service can monopolise supply of that service, raise prices above the competitive level and make excessive profits.[3] Monopolists also find it more profitable to sweat their existing assets rather than introduce innovations, so end users suffer additionally from that effect.

How likely is it that these adverse effects will materialise? The question can be answered both in relation to existing services and to services for which new licences are issued by the regulator.

The key question concerns the scope of market for spectrum. The demand for spectrum ultimately is derived from end users' demand for the services which rely on it and the ability of these services to make use of different frequency bands. The degree of interchangeability of

[3] Although some or all of these "excess profits" may be transferred to the government via auction proceeds.

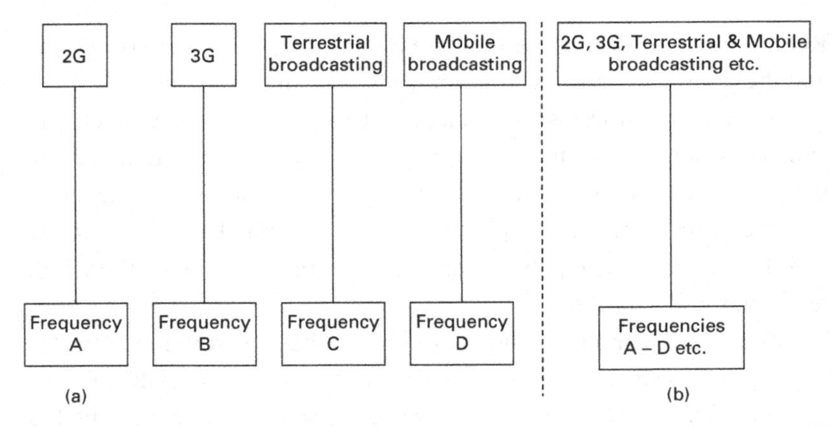

Figure 9.2. Various market scopes for frequencies.

different frequencies in providing services is determined partly by available technology and partly by the spectrum management regime (and in the longer term, the technologies which equipment manufacturers and operators develop and deploy depend upon what spectrum regulators allow).

Figure 9.2 shows two alternative arrangements. In Figure 9.2a, frequencies are allocated to particular services. In Figure 9.2b, the allocation exhibits service-neutrality: any frequency can be used for any service. Clearly, the task of monopolising spectrum is relatively easy under the former regime, but it may be impossible under the latter, especially if further frequencies E, F, G, etc., can be deployed to produce the same services. In other words, the risks of monopolisation are diminished by extending the frequencies over which a flexible choice of spectrum can be exercised.

As an example, in the UK's flexible spectrum market, the following frequencies are under consideration by operators for the provision of mobile broadcasting:

- digital audio broadcasting spectrum (at around 200 MHz),
- the so-called L-band (1452–1492 MHz),
- UHF spectrum, due to be released after the digital switchover (in the region of 700–900 MHz).

With such a choice of options, it would be very difficult for an operator to corner the market in spectrum for mobile broadcasting.

The first lesson, then, to avoid competitive problems associated with existing spectrum allocation under a market framework is to ensure an extensive scope for market operation. This removes barriers to entry. However, further measures can be taken to avoid any adverse effects, via regulatory intervention.

The task here is to strike a balance between allowing abuses of market power and penalising firms which, through their own efforts and innovation, have established a strong market position, but are still subject to competitive threats. This is a traditional problem in the application of competition law and policy, which seeks both to protect consumers from abuses and to encourage competitive investments by firms. A competition policy analysis of a particular concentration of spectrum ownership can be broken down into three stages as set out in Table 9.1.

This suggests a systematic way forward to investigate potential problems in spectrum markets. The problems include: hoarding or under-utilisation of frequencies; refusal to supply spectrum to competitors – thus foreclosing their entry into services markets – or achieving the same aim by excessive pricing of spectrum; abuse of market power in the services market, via excessive pricing or other conduct.

Depending upon the specific legislative frameworks in particular jurisdictions, this approach can be applied:

- under competition law, in the course of an inquiry into alleged abusive conduct by an operator;
- under competition law, in the course of a forward-looking inquiry into the consequences of a merger between operators, or the acquisition of a spectrum licence by the holder of another;
- under specific legislation relating to spectrum management, which gives the spectrum regulator the right to scrutinise the change of ownership of a licence, either in advance, or to investigate it afterwards.

Table 9.1. *Stages in implementing a competition analysis of spectrum markets*

	Stage	Method
1.	Define the market – i.e. the set of frequencies which are interchangeable in the provision of services to defined end user markets.	Examine what frequencies can be deployed, and whether they can provide effective competition to one another in the provision of services.
2.	Does the operator in question (now, or after a merger) exercise a high level of market power[a] in the spectrum market identified?	Examine the share of spectrum in the identified market held by the operator. Is it high (e.g. > 40%) and durable? Can it/will it be challenged by the availability of other frequencies?
3.	Has the operator abused market power by refusing to sell or lease "hoarded" or underutilised spectrum, or by abusing market power in any services markets?	Collect data on the operator's conduct in the spectrum market and in the services markets.

[a]In the European Union, the test under competition law is whether the operator is dominant, in the sense that it can behave to an appreciable extent independently of its customers and competitors, and ultimately of consumers.

The precise *modus operandi* will depend upon special features of a country's legislation and upon the regulator's preferences. In the UK, competition law alone is applied in relation to service markets.[4] This decision was taken not to require pre-notification of spectrum transactions, as such regulation would be disproportionate and in any

[4] Under UK law, the acquisition of a spectrum licence on its own is not susceptible to investigation under regulations applying to mergers, because spectrum on its own is not an "undertaking". However, an investigation of abusive behaviour in a service market could lead to remedies which related to the spectrum market.

case the spectrum regulator could revisit the issue if it seemed necessary.

One concern expressed about spectrum markets is that as well as hoarding they will exhibit speculation, i.e. the acquisition of spectrum ahead of need, either by a firm for its own subsequent use, or by an intermediary which acquires the licence expecting its value to go up, thus enabling the speculator to achieve a capital gain.

Such conduct, if separated from market power, is not likely to harm consumers. Speculators are using their own knowledge of future market developments to make what might be risky bets on how demand will develop, ignorant speculators make mistaken bets and lose money. But a farsighted speculator making successful investments is not only making a profit but also ensuring that spectrum is available to meet a market need. Provided it is divorced from market power, such behaviour is likely to benefit end users.

9.4 Spectrum caps

A more direct way of controlling spectrum concentration is to apply so-called 'spectrum-caps' to individual operators. These prohibit an operator from acquiring more than a specified number of MHz in any band. This can be imposed in connection with the auction of licences and may also be maintained subsequently.

The previous sections have noted that (i) there was a risk in an auction process that a single buyer might buy all the spectrum available, if by doing so it can create a monopoly position in a particular service market, and that (ii) the assignment of licences by beauty contest or auctions under a command-and-control spectrum management regime can prevent that outcome.

This suggests the desirability of a market-based policy which puts limits on licence acquisition when awards are originally made, to prevent monopolisation, but relaxes those limits later to allow greater flexibility in spectrum use. As an example, a significant number of frequencies in the United States form a flexible spectrum market providing Commercial Mobile Radio Services (CMRS). A spectrum cap was instituted limiting the amount of broadband CMRS spectrum an entity could hold in any particular geographic area. This applied both to "legacy" new spectrum and to spectrum acquired at auction.

The aim was to prevent a "carve-up" of spectrum by one or a small number of licensees. But nonetheless a cap does inhibit behaviour in the market. It penalises efficient operators and protects inefficient ones. Accordingly the FCC raised the cap in November 2001 and decided to eliminate ("sunset") it in January 2003.

Spectrum caps operating at the time of an auction – specifying maximum MHz awards, restricting an operator to a single licence, or (by extension) precluding named operators from taking part in a spectrum award – are widespread, and some regulators in a market environment for spectrum regulation have extended them into the post-award period. They can, however, have detrimental long-term effects and tight caps should be eliminated if there is evidence of effective competition, either in the spectrum market itself, or in the market for services which those (and other) frequencies support.

9.5 Conclusions

This chapter has shown that concentration of spectrum holdings in the hands of one or a small number of operators in commercial markets can provide a means to monopolise service markets.

The extent of the risk depends on policies adopted at the time of a spectrum award, and the rules governing spectrum holdings thereafter. There are grounds for designing award rules in ways which will prevent any operator from immediately gaining market power in a services market. This applies to both command-and-control and market spectrum management regimes.

As shown in Chapter 4 spectrum markets can provide a degree of flexibility and responsiveness to changing market conditions which is beyond the reach of command-and-control methods. Moreover, intervention by the spectrum regulator or by a competition or communications services regulator can reduce the risk of market failure associated with abuse of power in spectrum markets. For these reasons, concerns about competition are not a sufficient basis for rejecting spectrum markets. Instruments to combat monopolisation of spectrum in a market context are available, and the alternative command-and-control methods can have equally deleterious effects on competition.

10 Band management

10.1 Introduction

The introduction of trading and liberalisation broadly allows market forces to shape the usage of spectrum through licence holders making decisions on their use and ownership according to their market assessment. However, there may be some cases where market forces do not act strongly because licence holders do not have sufficient knowledge of spectrum, only have a small holding, or where the ownership is very fragmented and hence the transaction costs are high.

One possible solution to these issues is the emergence of intermediaries known as band managers. These are organisations which would make a business of acquiring spectrum and then leasing it to end users. If they were able to do this more efficiently than the regulator then they might be able to operate profitably.

In this chapter we look at the different classes of band managers that might emerge, the key issues for their success, and build a simple business case to show the policy and economic conditions that would be necessary for their profitable operation. This is to allow a greater understanding of the likelihood of band managers emerging and the impact of on-going policy decisions.

The idea of deliberately creating band managers for particular frequencies has been the subject of debate at least since 1987, when it was proposed in the UK as a possible means of privatising spectrum assignment and market testing the provision of the service delivered [1]. There were two elements to the discussions that took place.

- The possible superiority of private vs. public frequency assignment.
- The advantages of spectrum users pooling their demands and thereby using spectrum more efficiently. This is most obviously

possible where individual users have non-synchronised needs, so that by pooling their demands they can take account of the law of averages and achieve very high levels of certainty of access while making significant economies in overall requirements, compared with a state where each had permanent access to specified frequencies.

In this chapter we focus upon the second aspect and in particular investigate the trade-off between the value of potential spectrum savings from pooling demands and the costs of coordinating access to frequencies by multiple users.

We also briefly consider the tension between profits from the economies of scale made available by band management and the risks to competition which might result.

10.2 Types of band manager

Many different forms of band manager might be envisaged. We have suggested the following list although it may be possible that other types will emerge over time.

- **Administrative**. This is an organisation to which the mechanics of various aspects of the licensing process could be outsourced from the regulator. For example, they might handle licence applications in the private mobile radio (PMR) band on behalf of the regulator. This is more of an outsourcing arrangement than devolved band management and is not considered further here.
- **Industry cooperative**. This is a not-for-profit venture organised by an industry that understands it gets better spectrum efficiency if it acts together. An example in the UK is the Joint Radio Company (JRC) which manages spectrum on behalf of the utilities. We consider these to a limited degree in this paper but are more concerned with profit-making organisations.

- **Location owner**. This might be an airport that acquires spectrum covering its geographical area and then offers shared spectrum, infrastructure or services. There is currently a mixed picture here, with airports often deploying shared infrastructure but rarely owning the entire spectrum. Nevertheless, this might change under the new spectrum management mechanisms.
- **Commercial – licensed**. This is the "classical" band manager. Here a commercial organisation acquires spectrum and leases parts of it for a fee to end users. It might be a simple extension of existing licensing, such as putting more fixed links into an existing band, or it might involve innovative new technologies such as software defined radio (SDR) or ultra-wideband[1] (UWB) to increase the efficiency of usage. The time horizons may be short, providing spectrum for *ad hoc* activities, or long, providing a near-permanent use of the spectrum.
- **Commercial – unrestricted access**. This is similar to the licensed approach, above, except that the band manager offers a "private commons", allowing end users unlimited free access to the spectrum for a consideration, such as an initial royalty payment. This is an interesting case. From a regulatory point of view, the spectrum remains licensed (to the band manager) whereas to the end user it has much of the appearance of unlicensed spectrum. Hence the term "private commons", but equally hence the consideration here rather than in later chapters on unlicensed spectrum.

Note that there is a difference between a band manager and speculator. The aim of the speculator is to ultimately sell spectrum at a profit. The aim of the band manager is to make an on-going business from leasing their spectrum. We discussed the role of speculators in Chapter 9 and do not return to it further here.

[1] In fact, UWB is not a promising candidate for this application because it is unlikely a band manager would be able to acquire a sufficiently wide band to accommodate it without competition problems being raised. We consider this in more detail later.

10.3 Fundamentals of band management

10.3.1 Band managers need to use the spectrum more efficiently

For any band manager to exist – commercial or cooperative – there must be a gain to offset the cost of establishing and operating the band manager. This gain might arise through the following.

- **Greater spectrum efficiency**. The band manager might be able to accommodate more users in the band than can be achieved through individual licences.
- **Differentiated service**. A band manager might offer, for example, better quality of service by accommodating fewer users or better managing the band. It might offer more flexibility for short term spectrum requirements than would otherwise be possible. Alternatively, the band manager may offer a lower quality of service at a lower price for those who prefer it.
- **Brokering**. A band manager might be able to perform a brokering or coordination service that the users could not do by themselves, e.g. bringing together two disparate users to share a band who would not have likely found each other and bid jointly on that band.

The first two of these rely on finding a more spectrum-efficient approach to using the band. This is most likely to be achieved through the use of technology. Brokering is more of a coordination function relying on databases or knowledge management tools.

The drivers for cooperative band managers may be somewhat different. They might provide the following.

- **More flexibility**. Their customers might need continuous changes to their assignments which the regulator would find overly burdensome to supervise. The only way to gain this flexibility might be to manage the band themselves.
- **Access to additional spectrum**. There are examples in the UK where a cooperative band manager also manages bands on behalf of

users not in the cooperative, so that they can have access to additional spectrum, which they can then pool with their existing holdings to obtain greater throughput.

10.3.2 Planning and technology are the most likely routes to greater efficiency

There are likely to be many ways in which a band manager might increase the efficiency with which the spectrum is used. These include the following.

- By pooling the demands of users whose need for spectrum is intermittent.
- By using better planning tools which allow the margins for error to be reduced. For example, current assignments by the regulator often err on the side of caution. By using a more accurate tool, or rapid problem resolution procedures, more users might be accommodated within the same band.
- By owning a larger amount of spectrum than any individual user and deploying an effective sharing system between users and therefore benefiting from the effect that a greater number of users can make more efficient use of the spectrum. This requires a relatively pro-active style of band management where sharing rules are defined and monitored. It may require the band manager to deploy radio equipment to control spectrum access. In this case, they would become more of an operator than a band manager.
- By making better use of technology. This may be as simple as imposing a consistent and efficient technology upon the users and benefiting from better coexistence between users. At the other extreme it might involve making use of new concepts such as better instantaneous access allowing "gaps" that appear in time to be filled by opportunistic transmissions (broadly, the cognitive radio approach). We discuss advanced technology in more detail in the section below.
- In the case of a commercial – unrestricted access band manager, by imposing better politeness rules in licence-exempt applications so

that more users can be packed in or the quality of service can be higher. This is something the regulator can do but the regulator may be slow to react, creating an opportunity for the band manager.

These methods are aimed at a moving target. Where bands are effectively managed by the regulators then they will be continually seeking to enhance the efficiency with which they manage spectrum in order to perform their statutory duty to achieve efficiency. If a band manager demonstrates the value of a technical advance, the regulator may subsequently adopt it in licensing policy. This adds an element of uncertainty to the band manager, particularly where the business case requires a long payback period. Where spectrum is traded, market incentives will operate generally to encourage efficiency in spectrum use, reducing the band manager's advantages.

10.3.3 Advanced technologies

In Chapters 2 and 3 UWB and cognitive radio were identified as potential approaches to more efficient uses of the spectrum. In this section we look at whether they might be suitable tools for a band manager to achieve efficiency gains.

UWB is unlikely to be viable for a band manager. UWB transmissions require a minimum of around 500 MHz of spectrum. It is unlikely that any band manager would be able to acquire such a broad part of the key spectrum bands where UWB is most likely to operate. If, instead, UWB is treated at a regulatory level as licence-exempt band managers would be unlikely to profit from this; they might instead see interference levels rise and so band capacity fall marginally.

On the face of it cognitive radio appears to offer more potential. However, cognitive radio can work effectively only if there is a set of transmitters providing information as to whether the spectrum is currently free. By definition, band managers would not typically own such a network themselves. If they did build such a network they would most likely become an operator because it would only be cost-effective to build such a system if it also carried user data. At this point, the band manager would look very similar to existing shared business radio networks.

Alternatively, band managers might make use of the networks of others, such as existing network operators, to distribute occupancy information. They would need to find an operator close by in frequency terms since otherwise the cost of the terminals would rise significantly. Finding such an operator, and paying the price that they might be charged, are likely to be a major impediment to a band manager deploying cognitive radios. Finally, cognitive terminals are not yet available and might be expensive when they emerge. For these reasons, we do not see cognitive radio as a likely technology for band managers until at least 2015, if at all.

Another approach is for a band manager to partition spectrum flexibly among a user group. To do this, they might insist on all users having the same technology, able to operate across the entire band owned by the band manager. Each user group would put up its own transmitters, which would provide information on the instantaneous availability of spectrum. If one user group wished temporarily to exceed its spectrum allocation it would be able to do so assuming other user groups were not fully utilising their allocations. Monitoring and billing arrangements, which might be quite complex, would be needed to ensure that no one group was consistently exceeding its allocated amount of spectrum. However, this is just a short step from a truly shared infrastructure. Providing such a shared service is likely to be less expensive and less complex than such an arrangement.

Hence, we conclude that neither UWB nor cognitive radio will provide an advantage to the band manager.

10.4 The business case for band management

10.4.1 Introduction

The emergence of band managers will depend predominantly on a viable business case. In order to understand this better, we have developed a simple business case model. This is not intended to provide a detailed and accurate assessment, but to illustrate the key factors and to understand how these can be affected by policy decisions. In building our business case we have consulted with existing band managers.

Some of the elements of the business case are relatively straightforward. We have been able to understand staff costs, accommodation and administrative costs, and the start-up costs of planning tools by comparison with similar existing organisations.

Some of the other elements are more difficult, namely the following.

- **Cost of spectrum**. There are two potential cost elements – the cost of acquiring spectrum and any on-going spectrum fees that might be payable. We have adopted a "best guess" of likely fees in our model. The acquisition cost is more problematic. Given the uncertainty in this area, we have treated the cost of spectrum as a variable and examined the sensitivity of the business case to it. Importantly, we have assumed that the band manager sells the spectrum at the end of a 10 year nominal business period and that the value of the spectrum appreciates in real terms at 5% per year during this period.
- **Efficiency gain**. As discussed in previous sections, the business case for the band manager relies on being able to accommodate more users than either the regulator or the users themselves could achieve. With uncertainty about current efficiency and the role of new technology, estimating efficiency gains is problematic. We have taken advice from existing band managers who believe that gains in the region of 50% are possible in bands which are currently shared, such as the business radio bands. These gains come from more intelligent and responsive packing of users into the spectrum. We have used this as a base for our modelling but conducted a sensitivity analysis. However, we do not expect to see significant efficiency gains across all bands. For example, the cellular bands are already effectively managed by the cellular operators who have strong incentives to enhance efficiency. We would not expect a band manager to be able to achieve sufficient gains in these bands to make the business case viable.

- **Speed of filling the band**. A band manager would likely take some time to acquire a customer base. During the period of growth they would have limited income to cover their start-up costs. They may adopt policies of limiting their spectrum acquisition to match their growing customer base, or only acquiring spectrum from users who would then become their customers. Regardless, there will be a period of time before the revenue stabilises. There is little experience as to what this might be. We have made varying assumptions about this in the different band manager cases, as discussed below.
- **Savings available to users, via a band manager**. Recourse to a band manager would free a spectrum user from the cost of getting a licence from the regulator, and of subsequent interactions with the regulator. We have not been able to get an estimate of this saving and assume it is zero.

We have looked at three different cases, those of mobile radio, fixed links and unlicensed band management, to see whether there is significant difference in the key factors. Each of these areas is discussed in more detail below.

10.4.2 Private mobile radio

Private mobile radio (PMR)[2] may be a suitable candidate for band management. With many users sharing the bands it might be possible to achieve efficiencies by optimising their usage and technology. The actual bands that would be suitable would require further study, but might, for example, be found in the region of 235–400 MHz. Our modelling suggests that the spectrum manager would realise a small positive NPV of the order of $8 million. This would become negative if efficiency gains fell below 40%, or spectrum costs rose even slightly. Profitability also requires a high spectrum fee. This is because the primary efficiency gain is realised by paying a fixed fee to the regulator

[2] Private mobile radio systems have many different names – for example they are sometimes called specialised mobile radio (SMR) in the USA. They are systems typically owned by the end users, for example, by a taxi company that uses the system to better run its business.

but by being able to gain revenue from additional users each of which can be charged a fee. A fee level of above $26 000 per 12.5 kHz national channel per year is needed to make the business profitable.

Because many costs are fixed, the more spectrum that is managed the greater the economies of scale. Our model suggests that the band manager would need to manage some 2×4 MHz of spectrum to be profitable.

Overall, the impression is of a somewhat fragile business case. If any of our key parameters deviate from their assumed value, the business case fails. Add in the need for a large up-front capital investment to acquire the spectrum and the uncertainty surrounding trading and the business case appears marginal.

If the regulator wished to encourage the emergence of band managers in this band then our model suggests that they would have to increase the price of spectrum, or refrain from implementing capacity enhancement mechanisms themselves. Neither is consistent with good spectrum management practice.

10.4.3 Fixed links

Fixed links are another area where there would appear to be significant scope for band management activities. Regulators have in the past devolved management of some of these bands to major users such as BT in the UK. Many thousands of fixed links exist, most of which are licensed individually. This results in bands effectively being shared, both geographically and in frequency terms.

There appears to be less publicly available information about fixed links than PMR, and so we have found it even more difficult to make assumptions about the key parameters. We are therefore less confident about our conclusions than for PMR, but nevertheless, put a "first pass" business case forwards as a basis for further discussion.

In overview, the business case for fixed links appears much more robust than for PMR. The cost of spectrum could rise significantly before seriously impacting the business case. The fee charged for each fixed link needs to be in excess of $300 per year in order to generate a profitable business case and the band manager needs to manage in excess of around 2×25 MHz for profitable operation.

10.4.4 Unrestricted access or "private commons"

The business case for an operator offering unrestricted access to licensed spectrum is less clear even than that for PMR and fixed links. As far as we are aware, there are not currently any band managers of this type. In principle, such a band manager would acquire some spectrum, perhaps through trading, and would then allow access to a particular type of equipment, or class of user. Users' access to the band would be relatively unregulated so that to the user it would appear that they were using licence-exempt spectrum.

A key question is how the band manager would generate revenue. Ideally, revenue should be linked to usage, but for many unlicensed devices the overhead of collecting usage information and sending it, perhaps via cellular technology, to a central billing system would overly complicate the unlicensed device. This might make it too expensive or bulky to be viable. A simpler, but less optimal, solution is to charge a "royalty" fee as a one-off payment made at the time that the device is purchased. This might even be collected from the manufacturer in much the same manner that royalty payments for intellectual property rights are collected today. We have adopted this approach in our modelling, but note that other approaches might be viable and indeed perhaps result in a better business case.

Modelling the business case shows that even our assumed level of 2.5 million users is insufficient to make a profitable business, predominantly as a result of the large start-up costs associated with acquiring spectrum, unless fees are raised to \$100 per user, and that spectrum needs to cost less than \$1 million/ MHz for the business case to be viable.

This makes clear the difficulty of making a viable business case in this area. We continue to emphasise the preliminary nature of our business cases, but conclude at this stage that band managers for unlicensed spectrum or private commons will probably not emerge.

10.4.5 Summary of cases

These three cases have resulted in quite different conclusions.

- PMR band manager: the business case looks poor with generally insufficient return on capital. This is because of the likely high cost of acquiring spectrum within these bands.
- Fixed link band manager: the business case here looks promising. This is because spectrum is less expensive in the higher frequency bands. Also, the current policy of charging on a per link basis is favourable to a band manager who can manage many thousands of links and hence generate sizable revenue.
- Band manager offering unrestricted access or "private commons": it is unclear exactly how this approach might be arranged. For the method we have modelled, the business case appears poor because of the limited opportunities of raising revenue for unlicensed usage.

In general, band managers are assisted by high prices for access to spectrum, provided they apply to everyone, because such prices increase the pay-off to the band manager's greater efficacy. Also, the more inefficient the regulator, the greater the scope for the band manager to enact efficiency savings. We would expect regulators to strive to reduce pricing and to become more efficient and hence, if anything, to reduce the likelihood of the emergence of band managers.

10.4.6 Band managers and competition issues

Accumulating a large amount of spectrum in a band manager's hands could lead to competition problems if it allowed the manager to restrict availability, and raise the price of spectrum to users. Alternatively, the manager could price discriminate, making spectrum available in relatively large amounts to low value users, and using the resulting "shortage" to raise price considerably to high value users – thereby raising profits overall, but less obviously than by maintaining spare capacity. Another variant of abuse of a dominant position would be for a band manager vertically integrated with downstream activities to discriminate against its competitors to whom it supplied spectrum.

An analysis of these issues hinges crucially upon market definition, which itself depends upon the availability of spectrum which competes, in the relevant uses, with that supplied by the band manager. Regulators should be aware of this danger. They should prevent a band manager from abusing a dominant position or structure rules for the secondary trading of spectrum which prevent a dominant position from being created.

10.5 Summary and conclusions

A small number of band managers currently exist to manage spectrum on behalf of their clients. At present, these bodies are generally not-for-profit organisations established with the agreement of the regulator. With the introduction of spectrum trading, it is possible that commercial band managers might appear on the scene as licensees, providing services to third parties. Understanding under what conditions they might appear may help shape spectrum management policy.

There are many potential types of commercial band manager. These include site owners, profit making organisations offering licensed access and profit making organisations offering unlicensed access. Under an environment where spectrum is tradable, the key to successful operation will be the ability to make use of the spectrum more efficiently than the regulator, or a single user, can. This allows the band manager to accommodate more users, or offer a better quality of service to clients. Alternatively, the band manager might facilitate short term leasing of spectrum with more flexibility than the regulator can provide.

Much discussion has taken place about the role of cognitive radios. This technology has been suggested as suitable for band management. However, we do not believe this to be the case because of the need to construct an infrastructure to signal to cognitive devices. We do believe that the band manager can effect efficiency gains, but more through detailed planning coupled with a more detailed understanding of the needs and interference tolerances of their clients than would be typical for a regulator.

We have modelled the business cases for three different band managers.

- PMR band manager: the business case looks poor with generally insufficient return on capital. This is because of the likely high cost of acquiring spectrum within these bands.
- Fixed link band manager: the business case here looks promising. This is because spectrum is less expensive in the higher frequency bands. Also, the current policy of charging on a per link basis is favourable to a band manager who can manage many thousands of links and hence generate sizable revenue.
- Unlicensed band manager: this is a novel type of band manager establishing a private commons. The business case for this approach appears poor because of the limited opportunities for raising revenue for unlicensed usage.

The viability of a band manager is enhanced by higher spectrum charges or prices, because these enhance the value of the efficiency savings a band manager can bring. This ceases to apply only when high prices choke off demand from spectrum users. The more inefficient the regime of individual licensing, the greater the scope for the band manager to enact efficiency savings. We would expect regulators to strive to reduce pricing and to become more efficient and hence, if anything, to reduce the likelihood of the emergence of band managers.

Given our views on the economics of band management, we do not expect to see a proliferation of band managers. If regulators wish to encourage band managers they may need to look into specific policy measures to do so. More likely, they will adopt a neutral view and allow the most efficient forms of arrangements to result through the operation of the market.

Reference

[1] "Deregulation of the Radio Spectrum in the UK", CSP International, HMSO, 1987.

III Regulation

11 Incentive based spectrum prices: theory

11.1 Introduction

In earlier chapters we have stated that there is a need for, and a benefit associated with, regulating radio spectrum use. In practice the costs of regulation are typically recovered through licence fees paid by radio spectrum users and hence there is a price associated with the use of licensed radio spectrum. For example, in the USA the FCC applies two types of fees – application fees and regulatory fees which cover the administrative cost of managing the use of spectrum, respectively. They may also serve to discourage the filing of frivolous applications. If set too high, however, fees can result in under-utilisation of the spectrum, while if set too low hoarding and congestion may arise.

The simple recovery of administrative costs via licence fees, while practised by almost every spectrum management agency around the world, fails to make use of one of the most powerful incentive mechanisms available to encourage more efficient use of radio spectrum. By varying licence fees in a suitable way, a spectrum manager can improve the economic and technical efficiency of spectrum management. The setting of incentive based prices is especially attractive in circumstances where spectrum has been assigned and/or allocated via administrative means rather than auctions. Incentive based pricing also works well in the absence of secondary trading, but as we show in this chapter, it can also work alongside spectrum trading.

Licence fees are a potent means of achieving greater efficiency for radio spectrum licensees holding non-auctioned spectrum. Efficiency will often require annual licences fees to be set above administrative cost to reflect a range of spectrum management objectives (efficient management and use, economic, innovation and competition) having regard in particular to the expected future demand for spectrum. The application of incentive based prices for radio spectrum licences has been termed Administrative Incentive Pricing (AIP).

In this chapter we describe how the setting of incentive based radio spectrum licence fees or AIP can be undertaken to promote efficient use of spectrum in ways that go beyond the recovery of administration costs. A few countries have adopted pricing frameworks that are intended explicitly to promote economic efficiency. For example, in the UK Ofcom applies AIP. The application of AIP is based on opportunity cost principles which promote efficient use of radio spectrum. In a review of spectrum management policy in 2002 commissioned by the UK government the following was stated.

"The fundamental mechanism by which the spectrum management regime could contribute to economic growth is through ensuring that users face continuing incentives towards more productive use of this resource. The review considers that these incentives should be financial and based on the *opportunity cost* of spectrum use. In this way, spectrum would be costed as any other input into the production process. Price signals about the cost of using spectrum would be disseminated throughout the economy. This information should enable dispersed economic agents to make their own judgments about their use of spectrum and the alternatives open to them to meet their organisational goals." Opportunity cost is defined as: "the value of an asset or resource in the next best alternative that is foregone by virtue of its actual use." [1].

Economists have shown that where firms buy inputs on competitive markets this tends to promote efficient use of these inputs and ensures that outputs are produced at the lowest possible cost. In other words, when firms face the right incentives they choose inputs carefully to ensure that their operational costs are kept as low as possible. When firms use inputs like radio spectrum efficiently, economists describe this as a situation of productive efficiency.

The chapter is structured as follows. Section 11.2 looks at economic efficiency and radio spectrum, and Section 11.3 focuses on the productive element of this efficiency. Section 11.4 examines how pricing radio spectrum can be done to promote efficiency. In Section 11.5 the Smith–NERA method of calculating spectrum prices is discussed; this is an algorithm based on the application of the economic opportunity cost principle. Section 11.6 outlines how the Smith–NERA method might work in practice, and in Section 11.7 the parallel application of spectrum pricing and spectrum trading is discussed. Section 11.8 concludes.

11.2 Economic efficiency and radio spectrum

Many spectrum management authorities are required to promote efficient spectrum assignment policy. In the USA one of the core principles of effective spectrum management governing the operations of the FCC and the NTIA (National Telecommunications and Information Administration) is to maximise the efficient use of radio spectrum. In the UK, for example, the Communications Act 2003 states that spectrum assignment policy should ensure:

"the efficient use in the United Kingdom of the electro-magnetic spectrum for wireless telegraphy" [2]

Efficiency in this context is usually understood to mean *economic efficiency*: the attainment of outcomes consistent with the application of what economists call the Pareto criterion. The Pareto criterion asserts that economic activity is efficient when it is not possible to find an alternative way of undertaking the activity to improve the well-being of one individual without harming the well-being of at least one other individual.

Economic efficiency has three dimensions relating to production, consumption and the use of resources over time.

- Productive efficiency – production of goods and services ought to be undertaken at the lowest possible cost (cost is measured in terms of inputs). Users of radio spectrum should choose inputs, capital, labour and spectrum such that production of services is at the lowest cost.

- Allocative efficiency – the mix of goods and services produced must be optimal in the sense that no other mix can increase the well-being of one economic agent without harming the well-being of another economic agent. Spectrum should be allocated across different uses in a way that results in the Pareto criterion being satisfied.
- Dynamic efficiency – resources are deployed in a way that encourages the most desirable level of research and development and innovation. The use of radio spectrum should allow for the right amount of innovation.

Economic analysis has identified the structural and behavioural characteristics that are needed in an economy if efficiency is to be achieved. When all the conditions are met to ensure efficiency, this is called "perfect competition". In a perfectly competitive economy demand and supply for different goods and services, including radio spectrum, are brought into equality by the workings of the price mechanism and prices reflect opportunity costs.

In the equilibrium of a perfectly competitive economy the price mechanism establishes relative prices such that the cost to society of producing good X (say cellular telephony services) in terms of good Y (say broadband services) reflects consumers' willingness to pay for such a transformation (the opportunity cost). This result is known as the First Fundamental Theorem of Welfare Economics, and it is often used to lend support to the claim that competitive markets are desirable.

The insight that equilibrium prices are consistent with economic efficiency in a perfectly competitive market economy is a useful guide for spectrum pricing policy. In the context of spectrum management, it suggests that choosing prices for spectrum which equate supply with demand is likely to be consistent with efficiency.

However, it is also well known that in a perfectly competitive economy in which externalities feature (that is where the actions of some firms may impact on other firms) efficiency may be sacrificed – particularly where property rights may be difficult to define. As unregulated use of radio spectrum would give rise to too many interference externalities, we cannot rely solely upon the market as a mechanism for achieving efficiency. Nevertheless, a spectrum manager can use market incentives, such as AIP and spectrum trading, to achieve superior outcomes than alternatives such as command and control.

11.3 Productive efficiency and radio spectrum

In this section we apply a simple example to examine the relationship between the use of spectrum and productive efficiency which allows us to identify the conditions that need to be satisfied for productive efficiency in spectrum use, which in turn is useful when assessing spectrum prices.

Assume the available spectrum lies on a line between zero and one (the unit interval [0,1]) and that it can be used in two sectors 1 and 2 in the economy.[1] The sectors could represent broadcasting and telephony. To produce the final outputs in each sector, firms choose a mix of labour l and spectrum s – that is labour and spectrum are substitute goods.[2] We suppose that firms within a sector are identical. The total amount of labour is fixed and equal to $L = l_1 + l_2$, where l_1 is the labour used in sector 1 and l_2 is the amount of labour used in sector 2, and the price of labour, the wage rate W, is assumed to be determined on a competitive market. We assume also that the prices of all final outputs produced in the economy are determined in competitive markets. Spectrum, however, is allocated to each sector via an administrative process rather than via a market, and the price is assumed to be zero, with $\bar{S}$ the amount of spectrum allocated to sector 1; and $1 - \bar{S}$ is allocated to sector 2.[3]

Each firm using radio spectrum in each sector seeks to maximise profits and chooses an output level (and hence inputs labour and spectrum) to achieve this objective. Note that the firms face a spectrum constraint. At the allocation $\bar{S}$ total output produced in sector 1 is denoted $Q_1(\bar{S}, l_1)$ and output in sector 2 is $Q_2(1 - \bar{S}, l_2)$.

[1] Alternatively, consider the assignment of frequencies within a band used by two users.
[2] Assume that many other sectors exist in the economy, but these do not use spectrum as an input. However, the other sectors make use of labour and other inputs such as capital.
[3] For simplicity assume spectrum management costs are recovered by general taxation.

Given an administrative allocation $\bar{S}$, there are three possible scenarios with regard to spectrum:

(1) demand for spectrum in each sector is equal to spectrum supply in each sector;
(2) demand for spectrum is no greater than spectrum supply in each sector; and
(3) demand for spectrum in one or both sectors is greater than spectrum supply in one or both sectors.

From a policy perspective the interesting scenario is (3), where demand for (free) spectrum exceeds the fixed spectrum supply in one or both sectors. As the quantity of spectrum is fixed and finite, excess demand in one or both sectors raises the issue of whether a re-allocation of spectrum could bring about a gain in efficiency. Alternatively, could a re-allocation of spectrum free up labour resources (the other input) without necessitating a reduction in the quantity of output produced in each sector? If the latter were possible, then the released labour resources imply that additional output could be produced and this indicates the initial allocation is inefficient.

To examine whether a re-allocation of spectrum could deliver efficiency gains (or whether the current use of spectrum is consistent with productive efficiency), consider the effect of a hypothetical small change in spectrum allocation (a re-allocation). Suppose that, for a small increase in spectrum $\Delta\bar{S}$ allocated to sector 1, the initial output level in the sector $Q_1(\bar{S}, l_1)$ can be produced using Δl_1 units less labour. There are implications for sector 2 of this re-allocation, there is less spectrum available but on the other hand more labour is available. If it is possible for output in sector 2 to increase as a result of this re-allocation of spectrum (and labour) it would indicate that the initial administrative allocation is inefficient.

Although spectrum does not command a price in this example, it is possible to calculate its value by looking at the prices of other goods traded on markets. By doing this we can determine what might be termed the implicit price of radio spectrum and this could guide us in the setting of administrative prices. In other words, we are seeking to determine the value of spectrum in terms of the value of other goods used in the economy.

This can be illustrated by noting that output in sector 1 can be kept constant after the radio spectrum re-allocation by lowering the amount of labour used in the sector by the amount Δl_1. Because labour is traded on a competitive market at a wage rate W, this implies that the value of the marginal spectrum can be measured by the wage rate times the change in labour necessary to maintain output constant: that is as $W\Delta l_1$. The same reasoning can be applied to sector 2, where the value of $\Delta\bar{S}$ is $W\Delta l_2$, when assessed at $Q_2(1-\bar{S}, l_2)$. Efficiency occurs when the opportunity costs of spectrum are equal across the two sectors.

The values $W\Delta l_1$ and $W\Delta l_2$ are estimates of the opportunity cost of spectrum. Where $\Delta\bar{S}$ is allocated to sector 1, the economy foregoes $W\Delta l_2$, the value of the input resources that would be saved by allocating $\Delta\bar{S}$ to sector 2 to maintain production at output level $Q_2(1-\bar{S}, l_2)$. If the marginal unit of spectrum $\Delta\bar{S}$ were allocated to sector 2, by analogous reasoning the economy foregoes a saving worth $W\Delta l_1$ in sector 1. The values $W\Delta l_1$ and $W\Delta l_2$ allow AIP to be calculated correctly but in practice require an understanding of the relationship between radio spectrum and other close substitute inputs.

The values $W\Delta l_1$ and $W\Delta l_2$ are referred to by economists as the *marginal benefits* of spectrum. It is common for marginal benefits also to decline, reflecting the presence of decreasing returns to scale for firms. In other words, as firms produce more output the effect of a unit of an input on production falls. In Figure 11.1 the marginal benefits of spectrum are shown as declining in each sector. Sector 1's marginal benefit function is on the left and sector 2's marginal benefit is on the right hand side in the figure. As the amount of spectrum allocated to sector 1 increases, the marginal benefit of spectrum in sector 1 declines, whereas the marginal benefit in sector 2 increases.

In Figure 11.1 it can be seen that at the initial administrative allocation $\bar{s}$ the marginal benefit of spectrum in sector 1 is greater than that in sector 2. As stated above, the allocation shown at $\bar{s}$ is inefficient because the marginal benefit values, which are estimates of the

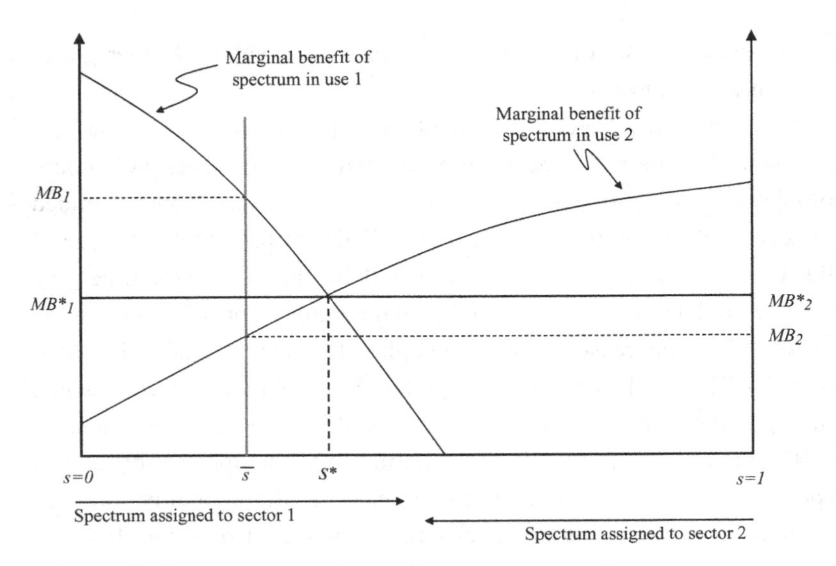

Figure 11.1. Marginal benefit of spectrum.

opportunity costs of spectrum, are not equal. Efficiency is satisfied at the allocation S^* in Figure 11.1.

It is very unlikely that a radio spectrum manager will be able to calculate S^* and achieve efficient spectrum management. The informational burden of assessing the correct allocation and assignment of radio spectrum administratively is enormous. Instead radio managers should, where possible and where desirable, look towards market mechanisms to help identify efficient allocations. In the following sections we address this by looking at how a radio spectrum manager can use estimates of opportunity costs to determine spectrum prices and use these to promote efficiency.

11.4 Pricing radio spectrum to achieve economic efficiency

The above suggests that to implement spectrum prices with efficiency in mind requires detailed information about the relationship between radio spectrum and other close substitute inputs. Alternatively, in the context of Figure 11.1 it requires a spectrum manager to have a good understanding about the shape of the marginal benefit functions.

While information about the shape of the marginal benefit functions is very useful, it is very demanding to expect a spectrum manager to be able to acquire all the necessary information easily. However, as we show in the next section, it is not necessary to know in detail the entire marginal benefit functions in order to compute spectrum prices that promote efficiency.

Prices that lead to improvements in the use of radio spectrum and shift allocations and assignments in the direction of efficiency can be devised using information about estimated marginal benefits at current allocations and assignments. One such method for calculating prices based on current assignments and allocations is known as the Smith–NERA method.

11.5 The Smith–NERA method of calculating spectrum prices

The Smith–NERA spectrum pricing methodology is a pricing algorithm used to calculate spectrum prices based upon opportunity costs [3]. We start by outlining a simple hypothetical example to illustrate the Smith–NERA method.

We assume that radio spectrum is in three non-overlapping frequency bands $\{a,b,c\}$ in the interval $[0,1]$. Further we assume there are three competing uses for this radio spectrum: I, II and III. We assume that a historical administrative allocation and assignment has occurred such that: Use I is allocated frequency band a, Use II is allocated frequency band b, and Use III is allocated frequency band c. The marginal benefits of the different frequency bands across the different uses are shown in Table 11.1. In addition, the marginal benefit of a non-spectrum input is also shown in the final column.

At the initial administrative assignment and allocation, spectrum in band a has a marginal benefit of 100 in Use I, whereas spectrum in band b has a lower marginal benefit of 75 in Use I and frequency band c has a zero marginal benefit in Use I. Thus frequency band a is the most

Table 11.1. *Marginal benefits of spectrum*

	Frequency bands			Alternative non-spectrum input
Uses	*a*	*b*	*c*	
I	**100**	75	0	0
II	35	**60**	30	0
III	10	10	**15**	5

valuable for Use I, and the frequency in band *c* has no value (that is, frequency band *c* cannot support the applications in Use I). The highlighted cells in Table 11.1 indicate the opportunity cost estimates that a radio spectrum manager could estimate at the current assignment and allocation – as data could be observed in the field to assist their computation.

The discussion on productive efficiency above suggests that marginal benefits across different uses for the same frequency band should be equalised if efficiency is to be achieved. The bold numbers in Table 11.1 show the case that the marginal benefits do not equal the values elsewhere in the same column (the same band across different uses) and therefore efficiency of spectrum use is not achieved. For example, frequency band *a* has a marginal benefit 100 to Use I, whereas in Use II it has a marginal benefit of 35 and in Use III a marginal benefit of 10. As all frequency band *a* is currently allocated to Use I, there can be no gain from re-allocating frequency band *a* to Use II as the spectrum is exhausted.

However, frequency band *b* has a higher value at the margin in Use I than it does in its current Use II. This suggests that by re-allocating spectrum in frequency band *b* away from Use II to Use I there is the potential for a gain in efficiency. As more of frequency band *b* is allocated to Use I the marginal benefit in Use I of using frequency band *b* will fall below 75, but also the marginal benefit of using frequency band *a* in Use I will fall below 100. Similarly, when frequency band *b* is taken away from Use II, the marginal benefit to Use II of frequency

Table 11.2. *Marginal benefits of spectrum following re-allocation of frequency band b*

	Frequency bands			Alternative non-spectrum input
Uses	*a*	*b*	*c*	
I	**90**	**70**	0	0
II	38	**70**	32	0
III	10	10	**15**	5

band *b* will increase above 60 and the marginal benefit for Use II of frequency band *c* will increase above 30.

The effect of shifting some frequency band *b* to Use I away from Use II leads to the revised marginal benefits shown in Table 11.2.

Table 11.2 differs from Table 11.1 in that the marginal benefits of frequency bands *a* and *b* in Uses I and II have changed. This reflects the fact that Use I has more spectrum and Use II less spectrum. It is also the case that the marginal benefit of frequency band *b* between Uses I and II is equal at 70. The equalisation of the marginal benefits indicates that frequency band *b* is allocated efficiently across uses. With Use II having less spectrum, the marginal benefit of frequency bands *a* and *c* in Use II have increased. The effect of re-allocating spectrum between Uses I and II also has a knock-on effect on the value associated with frequency band *c* in Use II.

In Table 11.2 above the bold numbers indicate those which a radio administrator may more easily be able to calculate in practice. It can be seen that there is scope for a further efficiency gains by re-allocating some of frequency band *c* to Use II from Use III. By doing this, however, the marginal benefit of frequency band *b* in Use II will fall (as total spectrum in the use increases). Table 11.3 presents the marginal benefits after re-allocation consistent with efficiency.

In Table 11.3 efficiency occurs where there is equality in marginal benefits across uses in the two highest values. It is not possible to find further re-allocations of spectrum that can yield better outcomes than

Table 11.3. *Marginal benefits of spectrum following re-allocation of frequency band c and further re-allocation of frequency band b*

	Frequency bands			Alternative non-spectrum input
Uses	*a*	*b*	*c*	
I	**87**	**68**	0	0
II	36	**68**	**25**	0
III	12	12	**25**	4

shown in Table 11.3 and therefore this is an efficient solution. It can be seen that to arrive at the efficient outcome the radio administrator needs to know about marginal benefits of frequency bands in neighbouring uses. Furthermore, the process of attaining efficiency is iterative – re-allocations are made one at a time (or several may occur simultaneously) and after each change the new marginal benefit values are assessed. The new marginal benefit values then inform the spectrum manager about the direction of further re-allocations or reassignments.

The next section discusses this iterative procedure in more detail and describes the least-cost-alternative method.

11.6 Setting spectrum prices to achieve efficiency using the Smith–NERA method

Tables 11.1–11.3 suggest an iterative approach is likely to be required in practice in order to achieve spectrum efficiency via administrative incentive pricing. Below we outline a framework for a radio spectrum manager that would support the implementation of an iterative procedure.

(1) Identify all frequency bands and associated uses (resulting in a matrix with a much larger dimension than the illustrative ones shown above in Tables 11.1–11.3). Populate the cells with estimates of the marginal benefits MB_{ij} appearing in use i (row) in

frequency band *j* (column) applying the least-cost-alternative. The least-cost-alternative is where a user substitutes spectrum with the least-cost-alternative input, such that output is unchanged following a small change in the amount of spectrum. For example, for a radio point to point fixed link the least-cost-alternative could be a fibre optic cable. In cellular telephony, the least-cost-alternative could be more or fewer base stations.

(2) Use the estimates of marginal benefits to identify the direction of change in spectrum re-allocation. For example, Table 11.1 above informed us that at the margin frequency band *b* is more valuable in Use I than in Use II. Hence there ought to be a movement of frequency band *b* away from Use II to Use I. Similarly frequency band *c* is more valuable in Use II than in Use III, so some frequency band *c* should move from Use III to Use II. The radio spectrum manager looks at the value of marginal benefits in a column (across the different uses) and seeks to move spectrum away from the use with the lowest marginal benefit to the use with the highest marginal benefit.

(3) Having identified the direction of re-allocation, which will depend on spectrum substitutability and marginal benefits, identify the maximum values of the MB in each column, for each column *j* call this MB^*_{ij}. In Table 11.1 these are 100, 75 and 30.

(4) If the maximum in step (3) occurs in a use which does not currently use the frequency band (such as Use I and frequency band *b* in Table 11.1), then spectrum prices should be set to lie in the interval between MB^*_{ij} and the current use marginal benefit. Hence, the price of frequency band *b* should lie between 75 and 60.

(5) Judgement is needed with respect to the actual price(s) chosen in the interval, but any information about the characteristics of the efficient allocation could guide price setting. Thus, if the values in Table 11.3 were known, this could inform the selection of prices. Assuming the data in Table 11.3 are not known, we propose that the price for frequency band *b*, for example, is set above 60 – but not significantly so.

(6) If the maximum in step (3) is the value for the current use of the band then set the price at this value.

(7) Having set prices for the spectrum, users will respond by changing their demands. After a period of time new marginal benefit values will emerge and the above procedure can be repeated. This may take up to five years or so. Eventually the radio manager will converge towards an efficient allocation of assignment of radio spectrum.

11.6.1 Setting the spectrum price: using judgement

Step (5) in the iterative process outlined above presents a question: if an interval is observed in marginal benefit values, what point or points in the interval should form the spectrum price(s)? Clearly setting prices too high will lead to a fall in the use of a frequency band and new demand would likely be small. This is clearly inefficient as spectrum would not be used. It is better that spectrum is used and contributing to welfare, than not being used at all.

Setting a price close to, but not equal to, the lower limit, would result in new demand for a frequency band, such as frequency band *b* in Use I, but existing demand by Use II would not fall by much. However, a price above the marginal benefit in Use II, but below Use I for frequency band *b* would lead to some spectrum being relinquished.

This approach in judging the right price for a frequency band serves to illustrate a more general point. Erring on the side of caution and approaching what the economists term the socially optimal price (*s*) (resulting in the equalisation of marginal benefits) from below is better for welfare.

11.6.2 Comments

The above is intentionally simple for expositional reasons. In practice there are many different firms operating in a use within a frequency band. In practice some firms in a use will find the AIP price too high, and other firms will find the price lying below their marginal benefit values. For AIP to work well, the selection of the representative firm has to be undertaken carefully.

There may be several different sub-uses occupying a frequency band, reflecting different final markets (e.g. in a PMR band there may be taxi firms, utilities and couriers). This is likely to give rise to different estimates for the marginal benefit of spectrum in a given use. A single measure of the marginal benefit may be calculated by taking a weighted average, where the weights to use could be the amounts of frequency in the different sub-uses.

The above analysis also makes no distinction across geographic areas. However, this can be accommodated by looking at matrices for different regions. In some regions where excess demand occurs, opportunity costs will play a role in influencing prices, whereas in other regions this will not be an issue.

Time is not considered explicitly in the above example. Demands vary through time, and some future uses may not be known. Prices should therefore be periodically re-evaluated taking account of changes in demand and technology.

11.7 The interaction between spectrum pricing and spectrum trading

Having discussed the formulation of spectrum prices aimed at achieving efficiency, this section addresses how AIP may work alongside spectrum trading, which was discussed in Chapter 6. So far in this chapter we have considered AIP on the implicit assumption that they were applied in the absence of spectrum trading. In this section we consider any implications of relaxing this assumption.

Three questions are considered, namely the following.

- What if any implications would spectrum trading have for AIP?
- What if any implications would AIP have for spectrum trading?
- Does AIP offer any potential advantages alongside trading?

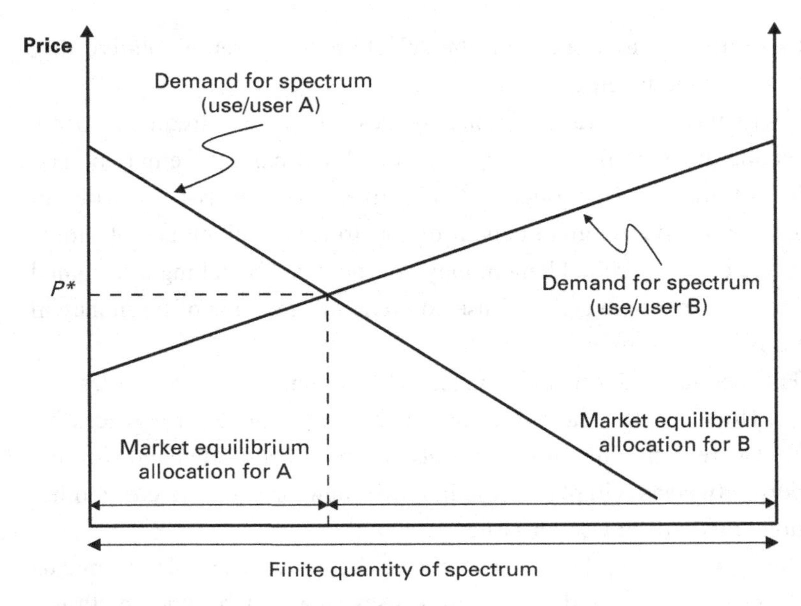

Figure 11.2. Spectrum trading and spectrum pricing.

11.7.1 Introducing trading

Figure 11.2 illustrates a situation where there are two uses/users of spectrum (A and B).

The horizontal axis is used to represent the finite quantity of spectrum available, and two demand curves are shown for competing uses/users (A and B) of the available spectrum. The demand curves reflect the private values of uses/users. If trading were allowed with AIP, then trading would occur if:

(i) AIP is set below market clearing prices (P^*);
(ii) the existing allocation of spectrum is inefficient;
(iii) transaction costs of trading (including any fees and taxes for trades) are less than the value of the potential gain from trade;
(iv) any information asymmetries do not prevent otherwise efficient trades.

Trading would therefore tend to lead to an efficient allocation in the absence of spectrum pricing, or where the AIP is set below the opportunity cost of spectrum. In terms of Figure 11.2 the competing uses/users would trade amounts such that the efficient allocation shown by the intersection of the two demand curves arises. For example, if B were initially allocated all the available spectrum, then A would purchase spectrum from B until the mid-point equilibrium were reached (subject to qualifications (i)–(iv) above). This is because A would be in a position to offer B a price above the value to B of the marginal spectrum.

11.7.2 Asymmetry of risk in setting AIP

If AIP is set too high this can give rise to unused spectrum which is very costly. If AIP is set too low then smaller losses corresponding to what economists call "deadweight loss" can arise. Since the latter are smaller than the former (for the same magnitude of error in setting AIP) AIP should be set conservatively, as discussed in the previous section.

With spectrum trading the efficiency costs from setting AIP too low will generally be smaller, and may be negligible, since spectrum users can trade and head towards the efficient outcome. The asymmetry in costs between setting spectrum prices too high versus too low where trading is permitted is therefore even greater, and the balance of risk therefore implies that AIP should be set even more conservatively.

11.7.3 Economic benefits and costs associated with trading and AIP

The benefits of efficiently set AIP and trading should in principle be the same – an efficient allocation of spectrum. This naturally raises the question of whether AIP adds anything in terms of efficiency.

In the absence of trading, and during a transition to trading, AIP can improve the efficiency of spectrum use. AIP may also have lower transaction costs, and potentially lower economic costs in terms of information asymmetry where the number of potential traders is very low (including the costs of information asymmetry). In particular, with a very small number of competing users of spectrum, efficient trades will not necessarily occur.

However, if market participants are uncertain about one another's demand for spectrum it is unlikely that a radio spectrum manager will have all the information necessary to set spectrum prices that would lead to efficiency. Nevertheless, AIP set on a conservative basis are likely to yield superior outcomes in circumstances where secondary markets are thin or non-existent.

AIP might also promote efficiency more effectively than trading where government or other not-for-profit entities are important spectrum users, since such agencies may be more responsive to an actual cost (with AIP) than an opportunity cost (with trading), as cost minimisation is likely to be an important objective for these entities.

11.7.4 Should observed traded prices inform AIP?

The analysis up to this point has left aside consideration of spectrum users' expectations regarding the introduction of spectrum pricing, and the response of spectrum prices to new information revealed by traded prices.

If AIP were linked to observed trading prices this could discourage trade (or at least discourage transparent trades). Linking AIP to observed traded prices could also discourage innovations that raise the value of existing spectrum.

To minimise any disincentive effect on innovation, one option would be to commit only to using information from trading with a lag in setting administered prices – in practice this may result in any case from the necessary administrative lag in resetting AIP.

11.8 Conclusion

In this chapter we have considered the setting of spectrum prices largely from a conceptual viewpoint. We have shown that a spectrum management agency can use prices to achieve efficiency in spectrum use, and that generally this will lead to superior outcomes than pricing on the basis of cost recovery.

We discussed the economic underpinnings of efficient pricing of spectrum and illustrated how prices can be calculated. The Smith–NERA methodology was discussed at length, and we indicated how this algorithm can be deployed in practice. We also discussed how spectrum pricing can coexist alongside trading.

The application of AIP was shown to be directed at efficiency – it is an instrument (the selection of the spectrum price) aimed at a target (efficient use of radio spectrum). AIP alone may not adequately tackle concerns about interference between different uses or users. If interference is an issue, then AIP needs to run alongside administrative regulations that target this problem.

In the next chapter we describe in greater detail the application of AIP.

References

[1] M. Cave, "Review of Radio Spectrum Management, for Department of Trade and Industry and H M Treasury", March 2002.

[2] Communications Act 2003, para. 152 section 5, HMSO, 2003.

[3] Smith–NERA, "Study into the use of spectrum pricing", report for the Radiocommunications Agency by NERA and Smith System Engineering, April 1996. Available from http://www.ofcom.org.uk/static/archive/ra/topics/spectrum-price/documents/smith/smith1.htm.

12 Incentive based spectrum pricing: practicalities

12.1 Introduction

In 1993 the ITU recommended that in spectrum prices should follow a set of principles [1].

- All spectrum users should pay a charge.
- Non-discrimination – the spectrum charge should be calculated fairly, i.e. if two users are using the same amount of spectrum in the same way, both should pay the same charge.
- The spectrum charge should be proportionate to the amount of bandwidth used.
- The charges should reflect the spectrum's value to society, i.e. if need be, frequencies used for public services should be subject to lower charges.
- The cost of spectrum regulation should not be borne by the state.
- Spectrum users should be consulted about intended adjustments in spectrum charges.
- The pricing structure should be clear, transparent and comprehensive, without unnecessarily lengthening the licensing process.
- The pricing structure should reflect the scarcity of available spectrum and the level of demand for spectrum in different frequency bands.
- The spectrum charge should be calculated so as to recover the costs of spectrum regulation. Spectrum pricing should not seek to maximise revenue for the government.
- The ability to levy spectrum charges should be anchored in law.

As discussed in the previous chapter, the contemporary approach to the setting of incentive based spectrum prices places a greater emphasis on economic factors. While some of the principles above remain relevant to the setting of spectrum prices, the 1993 ITU recommendations contain contradictions. For example, it is generally not possible to set spectrum prices to reflect scarcity while at the same time recover only the administrative costs of regulating spectrum.

Based on the discussion in the previous chapter, the main guiding principle that should be adopted for spectrum pricing is efficiency. This means the charges for using spectrum should reflect the value of spectrum to society, as measured by opportunity costs. Setting spectrum prices based upon opportunity costs can lead to:

- charges which generate revenue in excess of the costs of regulating spectrum; and
- some users or uses not paying a fee if opportunity costs are zero or negligible.

The best way to achieve efficiency through setting spectrum prices is to apply AIP which is based on the principle of opportunity costs measured according to the least-cost-alternative. In the remainder of this chapter we discuss how AIP can be set in practice. We highlight the challenges of designing AIP, particularly with regard to information requirements.

12.2 Applying administrative incentive prices: some issues

To apply AIP requires the application of the Smith–NERA methodology outlined in the previous chapter. This means a spectrum manager needs to know the input alternatives for the current radio spectrum used by an application and should understand what quantum of these alternative inputs would substitute for the current radio spectrum used. The steps facing a radio spectrum manager in the process of establishing AIP are as follows.

(1) For a given frequency band identify current and other potential uses of the band.

(2) Calculate the opportunity costs of spectrum for the current use of the band and other uses. This is achieved by applying the least-cost-alternative method described in the previous chapter.
(3) If there is a use with an opportunity cost higher than the current use, then set the AIP between the two values, but towards the bottom end of the range of values.
(4) If there is no use with an opportunity cost higher than the current use of the band then set the AIP at the value for the current use.

In principle the calculation of AIP is straightforward, but in practice there are a number of challenges which we discuss below. Note that the application of AIP is predicated on the assumption that it is possible to re-allocate spectrum administratively from the current use to other potential uses. The feasibility of achieving this over the timescale for which prices are set, say five years, needs to be considered. In bands that are shared between different uses, re-allocation will be relatively straightforward. However, it may not be the case for other bands. If it is not likely to be feasible to re-allocate the spectrum in these timescales,[1] then the opportunity cost for the current use should be used to determine AIP.

12.2.1 Calculating opportunity costs

Opportunity costs are calculated by using the approach discussed in the previous chapter. The marginal value of spectrum is the additional cost (or cost saving) to an *average* or *reasonably efficient* user as a result of being denied access to a small amount of spectrum (or being given access to an additional small amount of spectrum). The additional cost (cost saving) depends on the application and should be calculated as the estimated minimum cost of the alternative actions facing the user. These alternatives may include:

[1] This may be because there is no equipment available for the new use in the given frequency band. Administrative processes for re-allocating spectrum tend to be slow, though where administrative processes are replaced by market mechanisms, such as trading and the auctioning of overlay licences, the re-allocation of spectrum may occur over a shorter time period.

- investing in more/less network infrastructure to achieve the same quantity and quality of output with less/more spectrum;
- adopting narrower bandwidth equipment;
- switching to an alternative band;
- switching to an alternative service (e.g. a public service rather than private communications) or technology (e.g. fibre or leased line rather than fixed radio link).

The value of opportunity costs calculated by a spectrum manager will differ between the cases where spectrum is taken away and where spectrum is increased. For a marginal reduction in spectrum, the calculation will overstate the "true" value of spectrum, whereas for a marginal increase the calculation understates the "true" value.[2] An average of the opportunity cost values obtained from an increase and a decrease in spectrum gives a reasonable approximation to the true value of spectrum.

In practice, however, it is not always feasible to estimate values for both increases and decreases in spectrum and so there may be a small bias in estimates if the spectrum manager relies on one or the other.

12.2.2 Assumptions

To calculate opportunity costs a spectrum manager faces a number of modelling challenges. In particular, the spectrum manager will have to make assumptions in relation to a number of key questions.

- What is the appropriate size of a marginal change in radio spectrum?
- What is meant by an average or reasonably efficient user? For example, should the average user be categorised in terms of network topology and characteristics of equipment used (e.g. age, bandwidth, power), given radio communication demands (e.g. local or national, traffic levels, service quality requirements) or some other metric?

[2] The difference in values arises because a profit maximising firm, when faced with a change in the quantity of spectrum (having less or more), would respond by changing the output produced. However, in the least-cost-alternative approach, it is assumed that output is unchanged. While unrealistic as an assumption, it enables a proper test for efficiency.

- What is the appropriate discount rate and discounting period?
- How should equipment maintenance costs be assessed?
- How does one reflect the maturity of existing networks in the calculations?

Each of these questions is considered below.

What is the appropriate size of a marginal change in radio spectrum?

The calculation of AIP should be based on an assessment of the marginal value of spectrum, where marginal is meant to be a small change in spectrum used. Therefore a marginal increase or decrease in spectrum should reflect the minimum amount that is likely to be of practical benefit to the user. For example, in the case of a cellular network this should take account of typical cellular re-use patterns.

The amount of spectrum that constitutes "marginal" will also differ by service. For example, for PMR services marginal spectrum is likely to be a 2×12.5 kHz channel, whereas for aeronautical communications it is larger at 25 kHz and for a cellular network or PAMR services it is the number of channels required to populate a single cell "cluster", taking account of typical planning parameters.

Thus the marginal amount of spectrum depends on the use considered.

What is meant by an average or reasonably efficient user?

Defining a reasonably efficient user for the purposes of calculating marginal values should be based on information held by the spectrum agency in its frequency tables, and from information gathered from secondary sources and from industry. In some cases (notably for fixed links), the relationship between costs and bandwidth is non-linear, as a high proportion of costs are fixed (i.e. are independent of bandwidth). Consequently, there is a wide variation in opportunity costs determined for individual link types. One approach is to identify opportunity costs for each main link type and from these determine a weighted average value reflecting the total amount of spectrum utilised by each link type.

Inevitably some degree of judgement is used in deriving the assumed user profiles.

What is the appropriate discount rate and discounting period?
Current costs need to be converted into annual recurring values, as there are long-term and short-term costs associated with varying radio spectrum. The spectrum manager will need to make assumptions about discount rates, though these will be informed significantly by government assessments of discount rates and by the assessment of the cost of capital in the industry or sector concerned.

How should equipment maintenance costs be assessed?
As radio spectrum is more often than not employed by capital intensive industries, maintenance costs can be significant. Hence maintenance costs can impact materially the assessments of opportunity costs. There is no straightforward way of dealing with maintenance costs, though guidance may be found from the depreciation rates used in company accounts. In the UK a very simplistic approach has been advocated, where per annum maintenance costs are assumed to be 12% of initial capital expenditures.

How does one reflect the maturity of existing networks in the calculations?
When the amount of spectrum is varied in a use, the way a user responds by changing inputs will often depend critically on the maturity of the technology and/or network deployed. For example, in the case of GSM spectrum in Europe the networks are largely mature as they are fully developed in terms of coverage. Therefore a marginal change in spectrum will only affect the capacity of the network in areas where at peak demand the network is congested. To calculate the opportunity costs in this case the spectrum manager needs to understand how the GSM network performs at peak demand, and needs to understand the amount of infrastructure (i.e. base stations) that would substitute for spectrum.

12.2.3 Congestion and area sterilised

Clearly the opportunity cost of spectrum for a user or use will be related to the spectrum denied to other users, and the costs will typically be higher the greater the bandwidth used and the wider the geographic area over which use is denied, i.e. the area sterilised by the service. The concept of area sterilised is appropriate for services such as mobile and broadcasting but works less well for fixed links where congestion at specific nodal sites is often the main constraint on spectrum use.

If national prices are calculated, then the opportunity costs obtained for a local frequency assignment, such as PMR or CBS, can be converted to a national value by multiplying the local value by the likely amount of frequency re-use. This approach implicitly assumes that spectrum use is congested at a national level. It is important to test whether this assumption holds or not when converting marginal values into AIP. If the assumption does not hold and there is excess demand for spectrum in some but not all locations, then the national value could be calculated as relevant multiples of the congested and non-congested values where the multiples depend on the extent of congestion. An alternative approach would be to have geographic de-averaging of spectrum prices.

For some services (e.g. PMR or fixed wireless access) it may be appropriate to apply weighting metrics such as population or the number of businesses within an area where spectrum is consumed as a proxy for the degree of congestion. In other cases (e.g. fixed links), congestion may be measured in terms of the actual level of use at specific locations.

12.3 Calculating AIP in practice: case study of fixed links in the UK

12.3.1 Introduction

AIP has been in use since 1998 in the UK, following the passage of the 1998 Wireless Telegraphy Act. The application of AIP since 1998 has evolved and generally become more sophisticated. The amount of

Table 12.1. *AIP fees in the UK 2004–6*

	AIP by sector in the UK £000	
Sector	2004/2005	2005/2006
1 Aeronautical	818	931
2 Amateur and Citizen's Band	1030	883
3 Broadcasting	2454	4001
4 Business Radio	15 187	11 838
5 Fixed Links	18 203	20 895
6 Maritime	1723	2031
7 Programme Making and Special Events	1145	1412
8 Public Wireless Networks	63 868	63 011
9 Science and Technology	112	745
10 Satellite	928	974
11 Ministry of Defence	24 314	55 398
Total	**132 168**	**164 094**

Source: Ofcom.

revenue that was collected in 2002 is shown in the Table 12.1. Initially the scope of AIP was limited to some major commercial uses, but over time it has been extended to cover spectrum used by the emergency services and the military.

12.3.2 Setting AIP for fixed links in the UK: a case study

Fixed link services are point to point radio based services and in the UK they are used primarily for infrastructure links for mobile telecommunications networks. Each fixed link is separately licensed and there are over 40 000 links in operation. Popular frequency bands in use are 7.5 GHz, 13 GHz, 23 GHz and 38 GHz. There is also increasing interest in the 55 GHz, 58 GHz and 65 GHz bands for very short fixed infrastructure and access links.

In an Indepen study commissioned by Ofcom (see reference [2]), the following were considered to be main alternatives to fixed links.

(1) Use of more spectrum-efficient technologies within the same frequency band, i.e. allowing less spectrum to be used to convey the same amount of data.
(2) Use of a higher frequency band where there is greater capacity and less likelihood of congestion.
(3) Use of a non-radio alternative (e.g. a leased line or fibre).

We illustrate the calculations of opportunity cost for fixed links in the UK by looking at (1) above, which examines different technologies using the same frequency bands.

For most fixed links applications there are competing technologies, differentiated in terms of adaptive modulation,[3] which can perform the necessary data conveyance. Higher modulation schemes generally result in a lower spectrum utilisation per unit of data conveyed by the link, but cause greater interference and therefore require greater separation from other co-channel links. Thus there is a trade-off. Lower modulation schemes are spectrally inefficient but cause less co-channel interference, higher modulation schemes are more spectrally efficient but generate more co-channel interference.

The first task facing a spectrum manager in assessing AIP for fixed links is to identify the capital costs for the different technologies which can perform the necessary data conveyance. In the UK the consultants Indepen relied upon information from several sources to assess the costs of three different modulation schemes (QPSK, 16 QAM and 128 QAM) for six different data rates ranging from 2 Mbps up to 155 Mbps. It was assumed that a spectrally more efficient modulation scheme utilises 75% of the spectrum utilised by the next best less efficient alternative. It was also assumed that over 15 years of the lifetime of the equipment, a

[3] Adaptive modulation is used in many digital communication networks (e.g. cable modems, DSL modems, CDMA, 3G, WiFi, WiMax and point-to-point fixed links). Common techniques include quadrature phase shift keying (QPSK) and quadrature amplitude modulation (QAM). These techniques can be used to increase capacity and speed in a network. Modulation is the process by which a carrier wave is able to carry the message or digital signal. There are three common methods: amplitude, frequency and phase key shifting.

Table 12.2. *Data on fixed links*

Link speed and less efficient scheme	More efficient option	Spectrum utilisation		Equipment costs (£)		Value per 2×1 MHz (£)	Annualised value (£ per 2×1 MHz)
		Less efficient	More efficient	Less efficient	More efficient		
8 Mbps/QSPK	16 QAM	7 MHz	5.25 MHz	6 500	10 400	= £3900/1.75 = £2228	266

discount rate of 10% applied and that the maintenance costs of more efficient equipment was identical to that of less efficient equipment.[4]

The figures from the Indepen study for a link operating at 8 Mbps are shown in Table 12.2.

In Table 12.2 the value of spectrum is calculated by supposing that there is a marginal decrease in spectrum. For the operator to maintain data conveyance at 8 Mbps, a more efficient modulation scheme at 16 QAM would be the next best alternative and this would entail slightly less bandwidth (1.75 MHz). On the other hand the 16 QAM technology is more costly, £10 400 versus £6500. Hence, the value of 2 × 1 MHz is estimated by taking the additional costs associated with the more efficient technology and dividing this by the spectrum saved. Finally this figure is adjusted to obtain an annualised sum.

The computation illustrated in Table 12.2 shows that the estimated opportunity cost is sensitive to a number of variables: capital cost estimates, the perceived next best alternative, judgements about spectral efficiency gains, the discount rate (or cost of capital), and the lifetime of the equipment.

The Indepen study calculated values for a range of fixed link types and calculated a weighted average per MHz value based on the total bandwidth of each link type multiplied by the value per MHz and divided by the total bandwidth of all links in operation. The figure proposed by the consultants for the opportunity cost of fixed links was £132 per annum per 2 × 1 MHz per link. The latter estimate is sensitive to assessments about the bandwidth occupied by different link types and the number of links in operation across the different link types, as well as on the other factors referred to above.

The opportunity cost estimate proposed by Indepen for marginal spectrum used by a fixed link was reduced by the regulator Ofcom to £88 [3]. To price an individual link the reference spectrum price estimate is adjusted by a number of factors and the formula used by Ofcom is:

fixed link licence fee = spectrum price × as bandwidth factor × band factor × path length factor × availability factor.

[4] This is the value equivalent to the payment of a loan based on constant payments over a 15 year period for a constant interest rate (the discount rate) of 10%.

The spectrum price is £88 per 2×1 MHz bidirectional link and is calculated in the way described above.

The bandwidth factor takes account directly of the amount of bandwidth used by a link, for example in the 6 GHz band the average bandwidth is 37.37 MHz. Ofcom applies a minimum of 1 to the bandwidth factor.

The band factor reflects the balance in supply and demand on a band-by-band basis, and as such the level of congestion. The value is 1 for lower frequency bands and declines for links in higher frequency bands.

The path length factor reflects the opportunity cost of spectrum in a certain band, based on the extent to which shorter links deny spectrum to other users (of potentially longer and more efficient links) in that band. Ofcom operates a minimum path length (MPL) policy to conserve lower frequency bands for longer links which can be accommodated only in these bands. Whilst it is Ofcom's general policy to avoid making assignments where the link path length is less than the MPL, it does so when requested. When such assignments are made, the path length factor adjusts the fee by placing a premium on the use of path lengths below MPL. This premium reflects the opportunity cost of spectrum, based on the extent to which shorter links deny spectrum to other users in that band. For a given MPL for each band and system type, the path link factor is calculated according to the following formula.

When a link path length is at least as long as the minimum path length, the path length factor is equal to 1. When a link path length is less than the minimum path length, the path length factor equals the square root of the minimum path length divided by the path length. Ofcom caps the path length factor at 4.

Finally the availability factor determines the quality of spectrum a fixed link user receives (i.e. the probability that the fixed link user can receive a signal). A system availability requirement of 99.99% (sometimes referred to as "four nines" or "two nines") is the normal starting point when making assignments and is the most commonly requested value. However, other availability requirements are also available to suit customer needs. In developing the algorithm, the value of unity for the availability factor has been associated with the most common availability requirement (e.g. 99.99%). Higher (or lower) availability requirements attract a higher (or lower) availability factor, reflecting the opportunity cost of the spectrum denied to other users. The availability factor applied varies from 0.7 for 99.9% availability through to 1.4 for 99.999% availability.

12.4 Incentive based spectrum charges in other countries

Few countries to date have deployed incentive based spectrum prices based upon opportunity cost principles. Below we describe in brief those that have introduced AIP like prices or are considering doing so.

12.4.1 Australia

Australia has operated a system of spectrum pricing for a number of years that embodies in part the principle of opportunity costs. The fees are determined by the Australian Communications and Media Authority (ACMA), formed in 2005 from the Australian Communications Authority (ACA) and a broadcasting regulatory body.

The ACMA employs the following principles so that licence fees contribute to the efficient allocation of spectrum, and promote an equitable and consistent fee regime.

(1) Charges should cover the direct administrative costs of issue, renewal and instalment processing.
(2) Taxes from licensees as a group should recover the indirect costs of spectrum management (such as international coordination costs).
(3) Taxes should be based on the amount of spectrum denied to other users.
(4) Spectrum denied should be priced at its opportunity cost (the value of the best alternative use of that spectrum).
(5) If the opportunity cost is less than the indirect costs attributable to the licensee, taxes should only recover costs.

As in the UK, adjustment factors are applied to meet specific conditions.

12.4.2 Canada and Denmark

In Canada the government through the Minister of Industry has exclusive spectrum management responsibility of radio spectrum. Day to day spectrum management is performed by Industry Canada, a federal government department reporting to the Minister of Industry. Spectrum allocation is largely harmonised with the USA.

The total cost of the spectrum management program run by Industry Canada is around CAN$61 million per year and Industry Canada's licence fee revenue derived from non-broadcast activities is CAN$209 million per year. The cost of managing broadcast spectrum is CAN$13 million per year and the CRTC (the broadcast regulator) raises licence fee revenue of CAN$101 million per year. The fees raised are much in excess of the administrative costs involved.

The setting of spectrum fees in Canada is based on the market value or a reasonable approximation thereof of the spectrum used. As discussed in the previous chapter, the market value of spectrum can be estimated by its opportunity cost.

Since 1996 fees in Canada have been based on the quantum of spectrum authorised in a defined geographic area, with population or households included as a variable. Industry Canada is considering the introduction of an AIP-like mechanism called Spectrum Efficiency Incentive Pricing.

Denmark's regulator Telestyrelsen is also moving towards a spectrum management regime that incorporates greater use of market mechanisms and spectrum licence fees have a factor which includes an opportunity cost element.

12.5 Conclusion

Incentive based spectrum pricing is a tool that spectrum managers can use to encourage efficient spectrum use. Charging annual fees for the holding of spectrum is one way in which the spectrum manager can encourage current and prospective holders to make the right decisions to ensure efficient use of the spectrum.

Any use of spectrum imposes an opportunity cost on society – the value foregone of alternative use. This is because spectrum is finite and use is exclusionary – the use of spectrum for one purpose precludes its use for another. Therefore all decisions affecting current and future spectrum use should be made with a full and accurate reflection of these opportunity costs, if those decisions are to lead to the socially optimal allocation of resources in the short and long term. If the opportunity costs of spectrum use are ignored or discounted, socially suboptimal decisions will be made. One of the best ways of ensuring that the opportunity costs of spectrum are fully and accurately reflected by decision makers is for those opportunity costs to be reflected in prices that have to be paid to hold spectrum.

This is the principle behind the use of AIP. The primary purpose in applying AIP is not, in general, to achieve any specific short-term change in the use of spectrum. Rather, the aim is to ensure that the holders of spectrum fully recognise the costs that their use imposes on society by holding spectrum (or seeking to acquire additional spectrum), when making decisions. Many holders of spectrum are not in a position to make rapid changes to their use of spectrum in response to the application of AIP, but note that in practically every case the holders of spectrum have opportunities to change their use of spectrum in the longer term.

The use of AIP is justified by the benefits that should materialise in the longer term, as better decisions are made in light of increased awareness and appreciation of the value of spectrum – better decisions that should lead to more efficient use of the spectrum. The UK regulator Ofcom cited some evidence of the success of AIP. Since 2003 significant amounts of spectrum have been returned to Ofcom for re-assignment, as a more or less direct result of AIP. Some 28 MHz of the more valuable spectrum below 3 GHz has been released by public and private sector users in response to AIP, as has 160 MHz of the second-tier spectrum in the range 3–10 GHz.

References

[1] ITU, "Spectrum Pricing Study", Communication Study Groups, ITU-R SM.2012, 1993.

[2] "An Economic Study To Review Spectrum Pricing", Indepen, Aegis Systems and Warwick Business School, February 2004, available at http://www.ofcom.org.uk/research/radiocomms/reports/independent_review/spectrum_pricing.pdf.

[3] Ofcom "Spectrum pricing: A statement on proposals for setting Wireless Telegraphy Act licence fees", 23 February 2005, London.

13 How the commons works

13.1 Introduction

The "commons" is a part of the spectrum where anyone can transmit without a licence. For that reason it is sometimes called licence-exempt or unlicensed spectrum.

Unlicensed spectrum was until recently of little interest. However, since the late 1990s it has been debated more widely. This has been caused by the following developments.

- Deployments of new technologies in the 2.4 GHz band, particularly of Wi-Fi, have been very successful, leading many to ask whether further unlicensed allocations would result in more innovation and deployments.
- The development of ultra-wideband (UWB) and the promise of cognitive radio have led some to question whether these technologies can overcome historical problems with unlicensed spectrum.

The debate around the role of unlicensed spectrum has been particularly intense in the USA, where the term "spectrum commons" has come to be used to advocate an approach where much more of the spectrum is unlicensed. Advocates have suggested concepts such as radios seeking temporarily unused spectrum, making short transmissions and then moving onto other unused bands. Some of these concepts extend beyond unlicensed spectrum and into property rights and many of these issues were discussed in Chapter 7.

There is little agreement as to the optimal relative amount of spectrum assigned to unlicensed as against licensed usage. There are also many hybrid suggestions. For example, it has been suggested that spectrum be unlicensed but users have to pay a fee to access it depending on the current level of congestion. Alternatively, it is suggested that all spectrum be licensed but that licence holders be able to create "private commons" allowing a form of unlicensed access which they charge for in some form.

By far the most important band in terms of economic value is that at 2.4 GHz. The reasons why this band has proved so valuable are:

- it is available world-wide as an unlicensed band,
- the band is relatively large (83 MHz wide in most countries),
- it falls within one of the preferred frequency bands, having a useful range and relatively low cost equipment.

The development of this band is mostly fortuitous and based on the fact that the resonant frequency of water molecules is 2.45 GHz. This makes the frequency optimal for many heating applications including microwave ovens. This resulted in interference, which suggested to regulators that the use of the bands should be unlicensed. The same issues apply world-wide, hence the nature of the allocation. There appears to be no other band below 100 GHz where a similar physical property has resulted in another world-wide allocation.

In future, as shown by the 5 GHz band allocation process, a widespread unlicensed allocation will likely require coordinated regulatory activity with all the inherent problems and risks involved.

13.2 The economics of the commons

Initially we assume known technologies and demands for spectrum-using services, and on these undoubtedly unrealistic assumptions establish when spectrum should be allocated to exclusive use and traded, and when it should be a commons. The assumptions are then relaxed.

13.2.1 Case 1: a single output

Suppose a frequency has been allocated to a single service, and the only issue is whether spectrum has to be rationed. Suppose initially that the service requires spectrum in fixed proportions and it is competitively supplied under conditions of constant returns to scale.

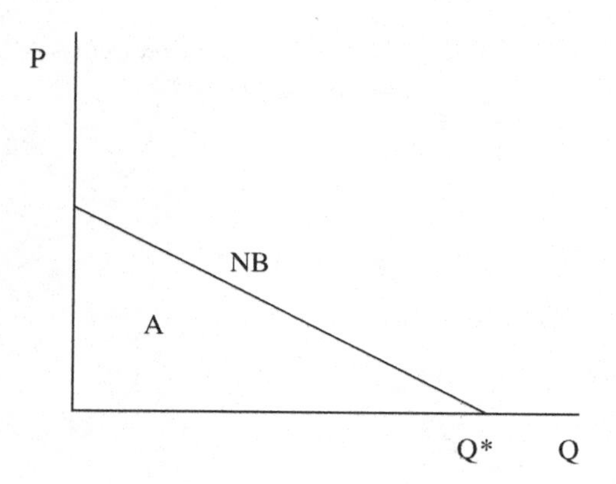

Figure 13.1. A commons.

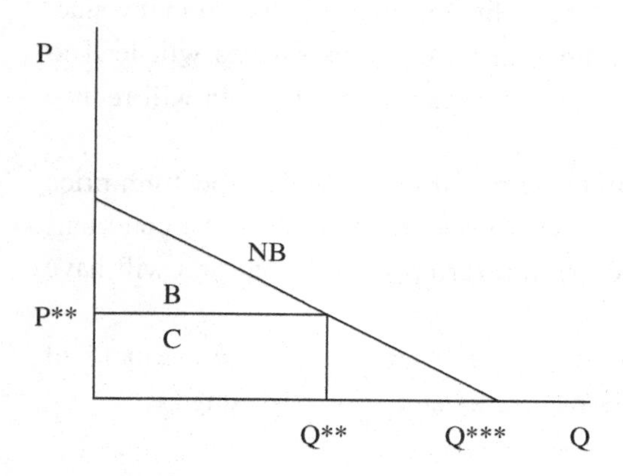

Figure 13.2. Scarcity.

The benefits of each regime are shown in Figures 13.1 and 13.2. The net benefit (NB) curve shows willingness to pay for the service minus marginal/average cost of production excluding spectrum. If enough spectrum is available to take output up to Q*, then the equilibrium price is zero. There is no need for rationing, and a commons, which avoids the administration cost of an exclusive licence and trading regime, is preferable. This is illustrated in Figure 13.1, where economic welfare is shown in the area under the net benefit curve (A), as this is the surplus over costs – or consumer surplus – which end users derive. In Figure 13.2, by contrast, spectrum constraints limit the allocation to Q**. The equilibrium price of spectrum is P**. At that price, economic welfare

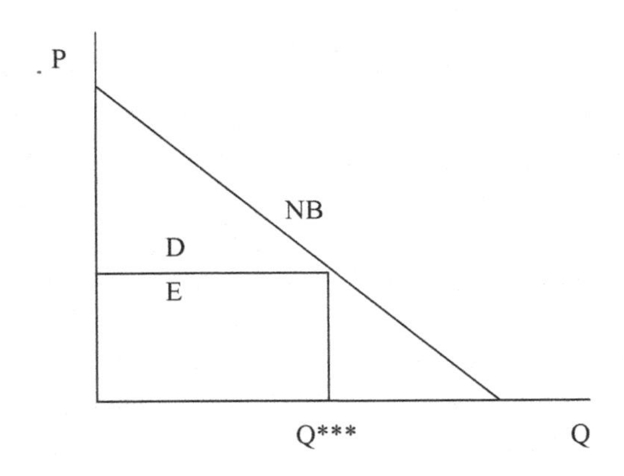

Figure 13.3. Demand from a competing industry.

is the sum of area B of consumer surplus and area C of spectrum revenue. Treating spectrum as a commons in these circumstances will lead to overuse, congestion and harmful interference, all of which will reduce the value of the service.

If other inputs can substitute for spectrum a positive spectrum price will provide incentive to cut back on spectrum use. Under a commons regime, however, firms faced with a zero price for spectrum will have no incentive to economise.

The results so far show that where there is an excess supply of spectrum, a commons works best. This is hardly surprising.

13.2.2 Case 2: competing outputs

Suppose now that a band can be used for two purposes, in one of which it is appropriately a commons, as in Figure 13.1. In the other, shown in Figure 13.3, there is excess demand for spectrum.

Which is the better use of the spectrum? Economic benefit in Figure 13.3 is shown by the sum of the areas D and E. Do they together exceed area A? In this case the areas are roughly equal, but clearly they need not be. If the net benefit curve in Figure 13.3 were much flatter, the alternative "commons" use would be preferable.

This shows that, if spectrum is intended to maximise economic benefit, conventional auctioning of exclusive licences is not enough. Either a prior decision has to be made to designate bands as a commons, or some public agency must estimate the benefits which would accrue to commons users, as a monetary sum, and be authorised to bid up to that amount in an auction.

13.2.3 Case 3: multiple bands

Suppose numerous bands are available simultaneously. In the traded sector, in the absence of allocation decisions, firms will sort themselves out through an auctioning or trading process to match services and frequencies optimally, but as in Case 2, commons may be undersupplied. If several commons exist, they may not be used with optimal technical efficiency, but this does not matter if there is no congestion in any band. One possible outcome would be for all spectrum prices to go to zero, permitting a universal commons, but this looks highly utopian in the foreseeable future.

13.2.4 Case 4: growing or uncertain demand

Reverting to Case 1, now suppose that demand for spectrum is currently less than supply, but this will change – or may change – in the future. If a commons is created, that will serve for now but the harmful consequences of congestion will or may occur later.

These effects can, of course, be mitigated by making more spectrum available for a commons later. If not, there seems little alternative to introducing property rights at the outset. Initially, the spot price for access to spectrum would be zero, but the asset price would reflect likely future scarcity.

13.2.5 Case 5: regulation of use of the commons

It is almost universal practice to postpone or avoid the effects of congestion by imposing limits on the purposes to which unlicensed spectrum can be put with respect to (i) use, including use to provide a service to the public, (ii) equipment permitted, (iii) the power at which the equipment may be used, and (iv) the enforcement of politeness protocols, which reduce interference.

13.2.6 Case 6: indirect revenue generation from unlicensed spectrum

Is it ever possible that a firm would bid for spectrum against other users and make it available as a "private commons"? It might be a viable strategy if another form of revenue were available – for example through the sale of equipment required to make the unlicensed use – although as discussed in Chapter 10 it is not clear that this will be the case. Analogously, before public broadcasting was developed, radio programmes were provided as a "commons" by equipment manufacturers, to increase sales of radios.

This is clearly a possibility, and the revenue available by this means would reflect the willingness to pay of consumers for the relevant final services. The outcome would not be perfect, as the owner of the private commons would have to restrict membership, to take account of the fact that members would tend to "overuse" the free (to them) spectrum.

13.2.7 Implementing the net benefit calculation

When a service is widely diffused, it is relatively straightforward to compute the net benefits. What we seek is the area under the demand curve, or willingness to pay, of subscribers. The net benefit is that magnitude, minus the non-spectrum costs of supply.

By way of illustration, we consider a paper on the demand for wireless Internet access in the United States [1]. This service uses licensed spectrum, but the same approach would be employed for a service which used unlicensed spectrum.

Like many telecommunications services, wireless Internet involves an access charge. A consumer will buy the service provided the surplus from usage (the amount by which willingness to pay exceeds the price) is greater than the access charge. This enables the investigator to estimate the distribution of consumer surplus from responses to questions about willingness to pay for access of the kind "what is the most you would be willing to pay on a monthly basis for wireless access to the Internet?".

Several other studies have used this method of establishing people's willingness to pay to estimate the value of spectrum. There is continuing debate about whether such statements about preferences are consistent with actual behaviour, but there is a growing consensus that, if appropriate questions are asked, the result can be relied on.

The difficulty, however, relates to new products. Respondents may genuinely be unable to express a willingness to pay for something which they do not understand. This would have restricted, for example, the application of this technique to WiMax prior to its introduction.

13.3 The likelihood of congestion in radio spectrum

13.3.1 Introduction

The previous section suggests that spectrum should be unlicensed where it is unlikely to be congested. This conclusion is widely supported by the findings summarised in Section 13.1. The logical next step is to determine the likelihood of congestion.

At present, congestion is generally defined as having more applicants for the spectrum then the available supply. Using this measure, congestion is dependent on frequency and location. It is also time-variant, growing in some bands, decreasing in others. However, many of the commentators have suggested that this is an inappropriate definition because those holding the spectrum may be using it inefficiently. This judgement is made on the basis of measurement activity [2] showing that some fully licensed bands have apparently little usage. Hence, they have concluded that if the licensing regime were changed to allow usage of the apparently unused spectrum then the pool of available spectrum would grow and the probability of congestion decrease. Easements and spectrum commons are both based on this logic. This conclusion, of course, assumes that usage would not grow to fill the space freed up by any such measures.[1]

[1] There is considerable complexity to understanding congestion. For example, a cellular operator can trade-off spectrum against cost by building a denser network of cells. This

In determining the most appropriate regulatory policy regarding unlicensed spectrum it is necessary to determine the following.

- Whether there is spectrum that is currently uncongested, can be expected to remain uncongested, and so could become unlicensed.
- Whether there is spectrum that is congested, but only because of inefficient usage and where changing the management policy to unlicensed usage would remove the congestion.

In this section we do not seek to perform a band-by-band analysis, but rather set out the principles by which such an analysis could be performed. Firstly, we set out the general causes of congestion in order to understand which bands in general might be less congested. Secondly, we address whether moving towards unlicensed usage will reduce congestion. Finally, we suggest mechanisms whereby congestion can be reduced in unlicensed bands.

13.3.2 Key factors which lead to congestion

There are many factors that influence congestion. Some of these are caused by suboptimal allocation policies and can be expected to be gradually alleviated by the introduction of trading. Others are caused by the nature of the radio spectrum. Very simplistically, frequencies below around 100 MHz have limited application because they propagate too far, preventing effective re-use. Frequencies above around 5 GHz are also less desirable because propagation is too short. In between these areas – and particularly in the bands around 500 MHz to 3 GHz – there is the greatest level of congestion.

There is nothing that the regulator can do to affect the relative desirability of these bands. However, there is one factor that the

requires less spectrum for the same capacity but costs more. Hence, their demand for spectrum will depend on the price. As the price falls, their demand will grow. If more spectrum became available, the price might be expected to fall and hence demand to grow until another point of equilibrium is reached. It might also appear that spectrum is congested at this point to observers.

regulator can control which has a significant effect on congestion. This is the maximum transmit power.

The shorter the range of transmission, the lower the probability that there will be two users in range of each other that might interfere. For example, at one extreme, a person using a garage door opener with a range of 20 m is highly unlikely to find another user of a similar device within the coverage area and operating their device simultaneously. At the other, in a cellular system with a cell covering a busy town, it is almost certain that there will be more than one person in the same cell transmitting at the same time during the peak hours. For a short range device with a maximum range of, say, 100 m, the coverage area, and hence the probability of congestion, is only 0.04% of a cellular phone with a range of 5 km.

Therefore, if only short range devices were allowed to use a particular piece of spectrum, the probability of congestion would be lower than for more general purpose spectrum. This would tend to favour unlicensed usage. Broadly, this has been the regulatory policy to date, with unlicensed spectrum having a maximum transmit power that tended to limit the range to around 100 m.

The other factors influencing congestion are the bandwidth and length of time of transmissions. These mostly depend on the usage. For example, the garage door opener needs to transmit only a short burst of narrowband data and on a only few occasions each day. A W-LAN base station might transmit broadband data almost continuously. The probability of congestion is proportional to this time–bandwidth product or information rate. Historically, most short range devices have also had a low information rate, but more recently W-LANs and BlueTooth have changed this trend. If the unlicensed bands were restricted to products with a low information rate then congestion would be lower. However, it is quite difficult for the regulator to restrict the information rate in an unlicensed band – the only feasible way to influence this is to ban equipment with a broad transmission bandwidth.

Hence, the main tool at the disposal of the regulator in controlling the level of congestion and the suitability for unlicensed use is the maximum transmit power, which equates to the range. By enforcing a low maximum transmit power, the probability of interference is reduced. Further, the amount of usage will also likely be reduced as some applications will not be viable with short range transmissions. Regulators might have a number of different bands with different transmit power limits to offer users different levels of range and congestion. Or, alternatively, as an unlicensed band became more heavily used the transmit power might be progressively reduced to new entrants in order to maintain the congestion at an acceptable level.

Mesh networks

Mesh networks have been proposed where the signal from a user is relayed by other users before eventually reaching a base station. It is not immediately obvious whether mesh networks qualify as short or long range communications systems since each "hop" might be only 100 m or so, but the overall distance between the user and the base station might be 1 km or more. In practice, they lie somewhere in between. A transmission of 1 km made by 10 hops of 100 m in length will result in a coverage area of 0.3 km^2. A single hop transmission of 1 km would result in a coverage area of 3 km^2. By comparison, a single 100 m transmission would have a coverage area of 0.03 km^2. In practice, it is hard to see how to prevent mesh usage in unlicensed spectrum since it will operate within the transmit power limits. Equally, mesh usage has proved difficult to realise to date, especially in a mobile environment, and it is far from clear that it will ever provide a cost-effective communications method.

13.3.3 Spectrum commons is unlikely to significantly alleviate congestion

Historically, the number of applications and users of radio spectrum has grown faster than the ability of technology to accommodate them. Hence, congestion has increased over time. However, it has been argued that if a "spectrum commons" approach was widely adopted then this would reduce the overall levels of congestion. This section considers whether this is likely.

It has been observed that, for much of the time, some of the spectrum apparently goes unused. This has led to the proposal that radios be allowed to locate and hop onto temporarily unused pieces of spectrum and remain there until the owner of the spectrum wishes to make a transmission. We discussed this concept in Chapter 2 where we noted that in its simplest form it would not work because of the hidden terminal problem and that as a result some form of central management was required to tell user devices whether the spectrum was free and to grant them access. Hence, we do not believe that such hopping behaviour, sometimes termed cognitive radio, will work without band management, which in turn implies some form of band ownership, rather than unlicensed use. In this section we explore the concept of band management further.

The basic concept of band management would be to create a large pool of spectrum. Owners of a piece of spectrum that was put into the pool might have guaranteed access to an equivalent amount of spectrum. They might also receive some payment for the additional usage that occurred on their spectrum.

Pooling of spectrum is effective under two conditions:

- individual holders of spectrum have insufficient spectrum to achieve good efficiency of usage, and
- different holders of spectrum have demand patterns that peak at different times.

Efficiency of use

The number of radio channels needed is generally calculated according to the Erlang formula. This shows that a few more radio channels are needed than the average demand would suggest in order to allow for demand peaks. The number of additional channels needed as a percentage of the overall number of channels falls as the total number of channels rises. This is shown in Figure 13.4, where the efficiency is compared with the number of channels for a 2% probability of blocking (a measure often used for cellular systems). The efficiency is the

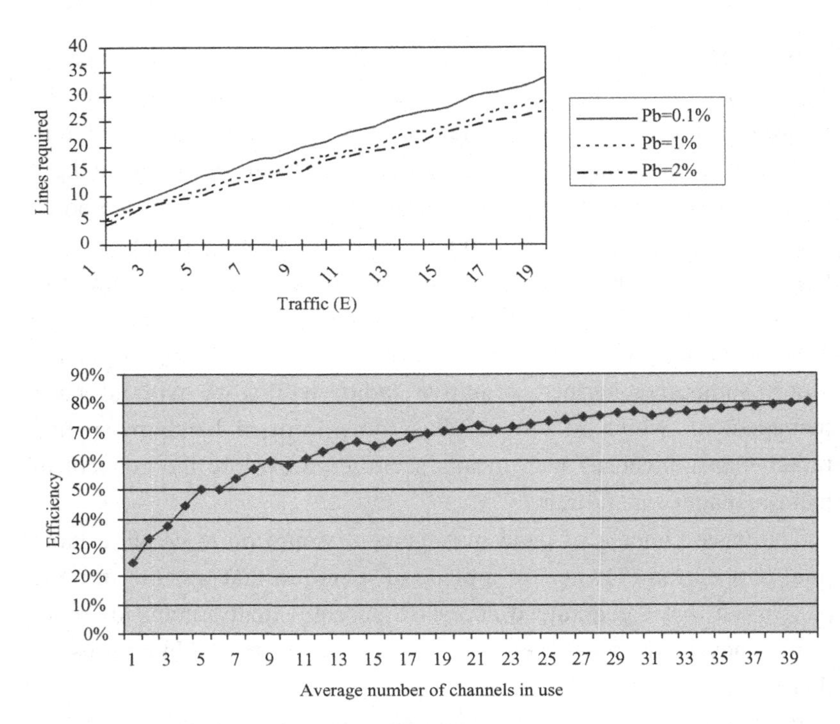

Figure 13.4. Efficiency of use of radio channels.

percentage of channels used on average compared with the total number of channels that needs to be set aside.

The slightly jagged nature of this chart is caused by the fact that there can be only integer numbers of radio channels. Although there is no clear point of inflection, the figure suggests that if the average number of channels in use falls below around 10, there might be significant efficiency gains from pooling channels with another operator.

In practice, most operators have many more channels than this. For example, GSM operators in the UK have around 1000 voice channels each. Even allowing for re-use across cells this is still more than 100 per sector. Hence, in general, we do not expect to see strong efficiency gains from pooling.

Differing demand patterns

If one operator had peak demand in the morning and another had peak demand in the evening then there would clearly be scope for improved efficiencies from sharing spectrum. The best way to understand whether demand patterns differ is to look at measurements of spectrum usage over time.

There are few published results of spectrum usage. The material here is taken from the FCC Spectrum Efficiency report [2]. This report made limited measurements and noted that many channels were lightly used, especially in the lower frequency bands (although not totally clear from the report, probably the bands below 500 MHz). The report also noted that public safety usage of radio channels is often below 10% utilisation but demand can rise to over 100% of capacity (i.e. blocking occurs) during peak usage periods corresponding to major incidents. The report noted that its measurement results were preliminary and that more work was needed but suggested that they indicated that pooling spectrum was likely to bring benefits.

The report itself notes that its measurement method will tend to understate the usage of the spectrum because of the following.

- A measurement made at a certain point might not be able to detect a nearby transmission if that transmission is behind a building. This becomes particularly difficult when CDMA technology is used as this results in relatively low signal strengths.
- Some operators use repeat patterns to avoid interference so that a particular frequency might not be in use in a cell, but might be used in neighbouring cells. This gives the impression of unused spectrum, but if the spectrum was used then interference might result.[2]

For all these reasons, measurements of usage are likely to return low results and need to be treated with some caution.

[2] This is a somewhat complicated issue. For example, it might be possible to use cellular transmissions in the unused TV transmitter bands because the cellular transmissions are at a much lower power level.

Even if the results are reasonably accurate, they suggest that spectrum is mostly under-utilised in the bands between around 200 MHz and 500 MHz. In the bands above this, corresponding to broadcasting and cellular, utilisation is much greater. This is unsurprising. In the 200–500 MHz band there is a mix of military use, public safety use and private use. Because much of the private use is not trunked it is known to be relatively inefficient. Also, private use is tending to decline somewhat year-on-year with the result that the spectrum is gradually becoming less used. The inefficient usage may change once spectrum trading is introduced. The economic incentives might result in more trunked usage and perhaps spectrum being used for different applications.

Even if there is under-utilised spectrum in this band, it is of limited value to the operators in the bands that are more congested. For example, these lower frequencies cannot be used efficiently by the cellular operators in congested areas because the propagation is extensive, tending to generate interference across multiple cells and hence provide very little additional capacity. They cannot readily be used for broadcasting since broadcasting requires quite broad bandwidth channels (e.g. 8 MHz wide for TV transmissions) which cannot be easily accommodated in the fragmented allocation pattern in the band.

It is also likely that much of the demand is correlated across operators. For example, all operators of cellular networks are likely to see similar demand patterns. Even emergency networks might see some correlation in the case where an incident causes disruption which in turn triggers mobile phone calls. Broadcasting networks have constant usage and so little advantage from pooling spectrum. Air traffic control, taxis, and other similar operators are likely to see peak usage around the same times as the cellular operators. Only, perhaps, military use might see uncoupled demand patterns.

There will be additional costs associated with pooling spectrum. User equipment will need to operate over a wider frequency range. Multiple control channels may be needed to inform the user equipment as to which frequencies are available but will themselves use spectrum and will require a constant stream of information to be passed between the various operating networks.

In summary, pooling of spectrum is a complex issue that merits a much deeper study. *Prima facie* the spectrum is likely to be more heavily utilised than measurements suggest and the difficulties in pooling are considerable. Hence, it remains unclear as to whether pooling is likely to be successful in reducing congestion.

13.3.4 In unlicensed bands, regulatory rules have an impact on the level of congestion

In addition to transmit power, there are some other rules which can impact the probability of congestion. These are:

- restricting the type of equipment which can be used – this will tend to prevent the band being used for certain applications;
- making the equipment more efficient so it uses less of the spectrum resource in transmitting its message;
- making the equipment "polite" so that it does not transmit if doing so would interrupt on-going transmissions.

The first approach essentially blocks a particular application from unlicensed spectrum, or from some of the unlicensed bands. Such a decision would need to be made on the basis that allowing this application would likely reduce the overall utility of the band. In practice, it would be an engineering-based judgement that allowing the application would result in a high probability of congestion, or excessive interference to existing users.

The last two approaches will tend to make the equipment more expensive for no apparent gain for the end user and so will require regulatory intervention. Even so, there may be enforcement problems, particularly if the increase in the price of the equipment is substantial. In this case users may be tempted to acquire simpler, non-approved equipment which might perhaps be legal in other countries. Because of the short-range and short-duration nature of most of the transmissions in these bands, enforcement could be difficult.

To date, the key regulatory mechanisms have been to restrict the equipment that can be used and to demand politeness. An extreme example of the former is the DECT band where only DECT equipment is allowed to operate. An example of the latter is the 5 GHz unlicensed band where European regulators have required that equipment using this band has dynamic frequency selection (DFS), which seeks a lightly used frequency within the band before transmitting.

Without any regulatory intervention there will be a tendency for none of these mechanisms to be used. Equipment will be made efficient or polite only to the extent that it is necessary for that piece of equipment to operate reliably and not for the greater good of all the users of the band. An example of this is BlueTooth and W-LAN in the 2.4 GHz unlicensed band. BlueTooth has been designed to use frequency hopping, which reduces the impact of interference on its operation. However, it also tends to increase the interference generated to systems that do not frequency hop like W-LAN. Studies have suggested that if both technologies operate in proximity then the BlueTooth system will work whereas the W-LAN system may stop functioning. Only with regulatory intervention has the BlueTooth standards body agreed to a modification in the standard whereby if a BlueTooth device senses the presence of a W-LAN transmitter it will not hop onto the frequencies currently being used by the W-LAN node.

By regulating the usage of the band the onset of congestion can be postponed but at the cost of increased equipment prices. From a theoretical point of view the optimal point is that at which the increased value of the usage of the band less the increased equipment cost is maximised. Practical factors related to enforcement also need to be factored in.

A further complication is that the increase in cost may depend on the device. For example, adding additional frequency hopping rules to a BlueTooth device which is already built around a complex integrated circuit will have minimal cost impact. Adding the same rules to a garage door opener which does not have sophisticated electronic circuitry could require a complete redesign with a much larger integrated circuit, significantly increasing the price. Equally, the potential for the garage door opener to generate interference is much less than the BlueTooth device because of the relatively infrequent usage of garage door openers. This might suggest different levels of regulatory intervention for different classes of devices, depending on the likely usage and cost increase.

By analogy, roads use some of these mechanisms to maximise their capacity. Certain types of vehicles are not allowed on roads, or are restricted to certain parts of the road – for example lorries are often not allowed in the outside lane of a motorway. Cars that are efficient in their use of space through being small are sometimes given tax incentives. Finally, politeness protocols are very widely applied, from deciding which side of the road to drive on through to regulating behaviour at traffic lights.

13.3.5 Conclusions

In the previous section we concluded that spectrum should be unlicensed if it was unlikely to be congested. As a result, in this section we considered the likelihood of congestion. We noted that:

- congestion was most likely in the core bands of around 100 MHz to 5 GHz;
- there was insufficient evidence that taking bands currently considered to be congested and making them unlicensed would alleviate congestion hence this approach cannot currently be advocated; and
- the probability of congestion could be dramatically reduced by restricting the range of devices through controlling the maximum transmitted power or by requiring specific behaviour such as politeness protocols.

However, there is no definitive way to predict congestion. A judgement needs to be made on the basis of the frequency band, likely use and range. The range in turn depends on the use. Hence, a key stage in predicting the congestion likely in the band is determining the most likely use. We discuss how this might be achieved in the next chapter.

13.4 Quasi-commons: UWB and cognitive radio

As mentioned at the start of this chapter, there are a number of approaches to commons which fall in between licensed spectrum and commons. UWB, if allowed, is unlicensed use in licensed spectrum. Effectively, in addition to setting aside the spectrum between frequencies A and B for commons, the regulator would also be setting aside power levels below C for "commons". UWB has already been discussed in Chapter 2. However, it is worth noting that if UWB is allowed it effectively opens up additional short range capacity, which might reduce the demand for "conventional" commons spectrum.

Cognitive radio also provides a form of unlicensed access into licensed spectrum. There are many possible variants of cognitive radio, for example:

- usage could be completely unlicensed with the only requirement being that the user must not cause interference to the licensed operation;
- usage could be semi-licensed with users having to register their equipment and perhaps needing equipment conforming to particular standards;
- usage could be fully licensed with users needing agreement from the licence holder before they can operate.

As discussed in Chapter 2, we do not believe that completely unlicensed use of cognitive radio will be possible without causing interference. Hence, we are more inclined to see cognitive radio as a variant of a licensed technology.

13.5 Summary

In this chapter we have argued the following points.

- An economic analysis of the situation suggests that spectrum should be unlicensed where there is little probability of congestion.

- Despite arguments about the ability of "spectrum commons" to alleviate congestion on a large scale, congestion across key parts of the spectrum is likely for the foreseeable future.
- Congestion is unlikely where short range communications are used and can be made less likely by regulatory insistence on such things as politeness protocols.
- Hence, there should be a mix of licensed and unlicensed spectrum, with the unlicensed approach restricted to bands and applications where congestion is unlikely.

In the next chapter we look at how a regulator might determine the most appropriate mix between licensed and unlicensed spectrum.

References

[1] P. Rappoport, J. Alleman and L. D. Taylor, "Demand for Wireless Technology: an empirical analysis", Presentation to the 31st Annual Telecommunication Policy Research Conference, September 2003.

[2] FCC Spectrum Policy Task Force, "Report of the Spectrum Efficiency Working Group", November 2002, http://www.fcc.gov/sptf/files/SEWGFinalReport_1.pdf.

14 Commons or non-commons?

14.1 Introduction

This chapter starts by considering whether market mechanisms could be used to determine the appropriate amount of spectrum commons. It then addresses two possible approaches that the regulator might use to make this decision, namely (1) the "total spectrum needed" approach and (2) the "band-by-band" approach.

14.2 The use of market mechanisms to determine the amount of spectrum commons

The standard market mechanisms are difficult to apply directly to unlicensed spectrum. Because there is no single user body, it is not possible for the unlicensed users to directly buy the spectrum under auction or trading.

A possibility is for a third party (the "unlicensed spectrum manager") to buy spectrum and to make it a private commons. Users wishing to access this would pay the unlicensed spectrum manager by some mechanism. The difficultly is in envisaging an appropriate mechanism. Ideally, the payment should reflect the level of usage, but for many unlicensed devices keeping account of the amount of usage and periodically returning the information would impose such a major increase in complexity and hence cost, that much of the revenue opportunity would be lost in subsidising devices.[1] An alternative is to impose a royalty-like fee on the manufacture of each device, with the

[1] Imagine, for example, the proverbial garage door opener. In order to account for usage it needs periodically to transmit to, say, a cellular network. It therefore needs a cellular transmitter and receiver with associated antennas and batteries. Most garage doors are supplied with multiple openers. The increase in cost, complexity and size would clearly be highly problematic.

fee level coarsely reflecting the expected usage (e.g. a garage door opener would pay less than a W-LAN node). This is more plausible as similar mechanisms are used today to collect royalty payments on patents.

This mechanism may work. We suggest that the auctioning and trading process be set up so as not to prevent it happening. However, we have concerns that the difficulties inherent in persuading manufacturers to accept such an approach are such that there is a possibility that potential unlicensed band managers will see the undertaking as overly risky. Further, our business case analysis set out in Chapter 10 suggested that the business case was very difficult for such a band manager. Hence, until such time as it emerges that the secondary band management is viable, we suggest that there also be regulatory intervention. This intervention can be periodically assessed to determine whether it is necessary and indeed whether it is hindering the emergence of market tools.

In the next two sections we look at two mechanisms that the regulator might use for determining whether the spectrum should be unlicensed.

14.3 The "total spectrum needed" approach

As we discussed in the previous chapter, unlicensed spectrum is predominantly viable for short range communications. Hence, one approach to understanding which spectrum should be unlicensed is to assess the total amount of spectrum that might be needed for short range communications in the future. This is speculative as the applications that might arise cannot be predicted with certainty, but nevertheless might provide a good indication of overall direction.

For example, it has been suggested that, at least for the next 10 years, 100 Mbits/s per person would be sufficient for short range communications. This would allow the transmission of multiple high-definition TV signals, audio streams and file transfers. Allowing for the frequencies to be re-used with a repeat pattern of around seven,[2] and for some inefficiencies in use of spectrum, this might result in an overall requirement in the region of 1 GHz. We should emphasise here that these calculations have only been provided by way of example, and detailed study would be necessary were a regulator to adopt this approach.[3]

Having made their prediction, the regulator would then need to identify spectrum that could meet the needs for short range communications and which could be freed up in a reasonable time period. Once the spectrum had been made available it could be monitored for usage, and the calculations of need periodically refreshed as applications emerged. Should it appear that the initial estimates were too low then the regulator could decide whether to make more spectrum available.

It would not be so simple in practice. Finding spectrum that can be freed up is always difficult and might become increasingly so if market forces such as spectrum trading were in use. Also, the regulator should decide whether it would be economically sensible to provide spectrum for unlicensed spectrum by considering whether there are other applications that would be more valuable to the country. However, the approach at least has the advantage of having a clear overall strategy.

14.4 The "band-by-band" approach

14.4.1 Introduction

An alternative to the "top-down" approach of estimating total spectrum need is instead to work from the bottom-up, looking at each band in turn. A possible bottom-up process is set out in Figure 14.1. We emphasise that, despite the use of a flow-chart, this is not an exact

[2] A repeat pattern of seven is classically used in cellular systems. Short range systems vary. The 802.11b standard only has three channels as a result of spectrum constraints and hence a maximum repeat pattern of three. DECT has ten channels. Further study to assess the optimum repeat pattern would be appropriate.

[3] In its "Spectrum Framework Review" the UK regulator, Ofcom, suggested that around 800 MHz of unlicensed spectrum might be needed on this basis.

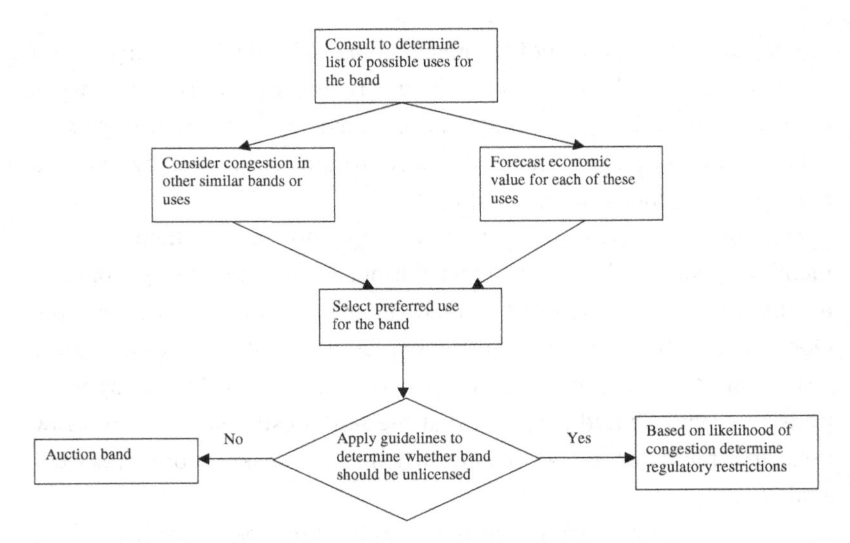

Figure 14.1. Outline of the process that a regulator might follow.

science and determining the correct answer is still likely to be difficult. In outline our process consists of four key stages:

- determine the most likely use of the band under licensed and unlicensed conditions;
- estimate the economic value generated under each of these uses;
- select the approach that maximises the economic value; and
- if unlicensed, determine which regulatory restrictions should apply.

Each of these stages is discussed in more detail below.

14.4.2 Consultation

Once a band had been identified as a potential candidate for unlicensed use, consultation could be used to gain an early understanding of likely use. Although potential users of unlicensed spectrum would probably not respond to the consultation, manufacturers of devices might. A specific question on the consultation document asking about unlicensed usage might yield important answers.

14.4.3 Economic assessment

The optimum allocation of spectrum would be the one that resulted in the greatest economic value for the country. In general, market-based methods of allocation and assignment will provide this outcome but, as we argued above, the market may not be able to reliably allocate unlicensed spectrum. An alternative is to attempt to predict the economic value of each of the different plausible uses and then to favour the use with the highest expected value.

Economic value assessment has been widely used in the UK. However, this assessment is normally retrospective, measuring the value of applications already in use. In making an allocation decision it is necessary to perform a forward-looking value assessment.

Forward-looking assessments follow the same methodology as retrospective assessments. The difference is that the key input data such as numbers of users and value of equipment needs to be forecast rather than observed. Hence, the major workload becomes one of forecasting. It is in this forecasting that the inherent problems with this approach reside. Forecasting future demand for wireless is notoriously difficult because of the changing applications and technologies in this area. For an economic assessment it might be necessary to forecast more than five years out. Because such forecasts have a low probability of being accurate, this economic valuation can only be seen as approximate. Were it likely to be highly accurate it would be the only input needed into the regulatory decision but because of the likely inaccuracy it should only be one of a number of inputs into the regulatory decision.

14.4.4 Examine similar bands and uses

The regulator might be able to learn from related occurrences in nearby bands. For example, if a potential use is fixed links, but a neighbouring band has already been allocated for fixed links and is under-used, then the use of the new band for fixed links might be given a downwards bias when compared to other uses. If usage is growing rapidly in unlicensed bands elsewhere in the spectrum then a bias towards making the band unlicensed might be appropriate.

14.4.5 Decide whether to make the band unlicensed

Based on the economic value comparison between licensed and unlicensed use modified by any relevant factors identified above, the regulator should be able to make a decision as to whether the band should be licensed or unlicensed.

14.4.6 An alternative

A variation on this approach can be adopted. Rather than estimating the likely value of the spectrum under both licensed and unlicensed conditions, instead the regulator could just estimate the likely value under unlicensed conditions. The spectrum could then be auctioned, with the value under unlicensed conditions set as the reserve price. If this price is exceeded, the spectrum is awarded as licensed spectrum to the highest bidder. If not, it becomes unlicensed.

While this has the advantage of the market estimating one of the values, there is a difficultly in that the value estimated by the regulator for the unlicensed spectrum is the economic value to the country whereas the price bid by potential licence holders is related to the producer surplus, or profit, they expect to receive from using the spectrum. However, given the inaccuracies in forward estimates of value, it may be that this difference is not overly problematic.

14.4.7 Deciding on regulatory restrictions

If the band is to be unlicensed then the regulator may wish to impose restrictions as discussed in Section 14.3. The regulator will need to make a judgement as to the most appropriate level of restriction.

In outline, the greater the perceived risk of congestion developing, the more restrictions should be imposed. However, the restrictions should also take into account the likely additional cost imposed on the devices compared to the benefit that might accrue. Depending on the level of information available, it might be possible to perform an economic assessment of the value of the different approaches.

For example, where imposing politeness protocols will have minimal impact on the device cost then they might be used without hesitation. Where such protocols would significantly increase the cost and where congestion is unlikely, or has little impact, then they should not be imposed.

The regulatory process is now complete. The process should be repeated periodically, and not just when spectrum is being auctioned as it might be appropriate as technology or applications change to acquire spectrum through trading and make it unlicensed.

14.5 Summary

In this chapter, we considered the role of unlicensed spectrum and in particular addressed the question as to how to determine whether there should be more or less unlicensed spectrum.

We suggested that it was unlikely that market mechanisms could be used in determining how much unlicensed spectrum would be needed so instead the regulator would need to make a judgement. We set out two approaches that the regulator could follow. The top-down approach estimated the total amount of unlicensed spectrum that might be needed and attempted to find sufficient spectrum to meet this need. The bottom-up approach considered bands of spectrum on a band-by-band basis, using a structured process to determine for each band whether licensed or unlicensed use would be most appropriate.

15 Is public sector spectrum management different?

15.1 Introduction

Spectrum is used to produce services which are supplied by firms for commercial reasons and distributed into a market place, and to provide public services such as defence and emergency services which are usually provided free at the point of delivery by a public body.

In the UK public sector spectrum use accounts for just under half of all spectrum use below 15 GHz – this represents the vast bulk of valuable frequencies. The breakdown of public sector spectrum use is shown in Figure 15.1.

In other countries too, military use of spectrum, particularly for radar and communications, accounts for most of public sector use. In the presence of international military alliances, such as NATO, military spectrum allocations are often harmonised internationally.

In most jurisdictions, commercial and public sector spectrum allocations are managed in a broadly similar way by the same independent agency or government department. A major exception is the United States, where spectrum used by the Federal Government is managed by the National Telecommunications and Information Administration (NTIA), part of the Department of Commerce, while spectrum allocated for commercial purposes and to state and local government is managed by the independent communications regulator, the Federal Communications Commission (FCC). Any major transfer or re-alignment of spectrum use may face the additional handicap of negotiations between these two organisations.

Historically, public sector organisations, especially national defence departments, were accorded high priority in spectrum use. Under the command-and-control regime, they were allocated spectrum for an

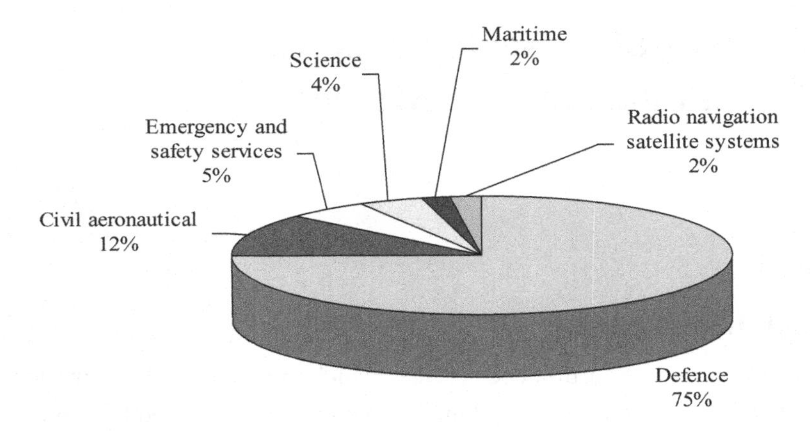

Figure 15.1. Breakdown of public sector spectrum.

indefinite period. This encouraged them to seek spectrum now, if they thought they might need it in the future. But as demand for commercial spectrum grew, attention became increasingly focussed on the issue of whether public sector bodies used spectrum inefficiently – in other words, whether they had appropriated too much.

In May 2003, President Bush signed an Executive Memorandum, requiring the NTIA and other federal departments to improve the efficiency of spectrum use. This includes the promotion of market-based economic instruments in spectrum management [1]. In the same year, the UK Government commissioned an independent audit of spectrum holdings, focussing particularly upon public sector spectrum use. In 2006, the Australian spectrum regulator commissioned a similar study. The purpose of an audit is to inquire whether there is scope for re-allocation from public to private sector or within the public sector. The next step, which this chapter discusses, is how to correct any misallocations. We address this issue by first asking whether the instruments of spectrum management discussed above can be applied to the public sector.

15.2 Is public sector spectrum special?

One of the themes of this book is that spectrum is one input among many in a variety of production processes. In a market economy inputs such as land, labour and capital equipment are distributed throughout the economy via a market process: the provider of capital or employee moves to whichever activity offers the best rewards. We also know that if markets in the economy for inputs and outputs are workably competitive, this system will promote economic efficiency, as inputs are put to use where they yield the highest returns. This logic underlies our discussion of spectrum trading in Chapter 4.

The question we now confront is: does this argument apply in the case of public sector trading? At first sight it may seem incongruous to require a public sector body such as a fire service or a defence force to compete in a market place for spectrum with commercial providers of services such as mobile broadcasting. However, this is exactly how public sector organisations acquire other inputs – such as employees, vehicles, land and office space. In relation to these inputs (with the exception of a compulsory military draft in the case of labour) public sector bodies have to go into the market, for example buying or selling land, hiring workers or leasing buildings.

The argument for special arrangements for spectrum for the public sector seems to be that (a) it is indispensable to the provision of service such as defence radar; (b) the service itself (such as an ambulance service) has a very high priority; and (c) under past spectrum management practice, the only way to acquire spectrum was by administrative methods.

The use of markets to allocate other equally indispensable inputs into vital public services appears to negate propositions (a) and (b) above, and (c) could be resolved by the development of a spectrum market place. In order to meet the needs of public sector users, the market would have to cover a large number of frequencies (to permit substitution) and trading opportunities would have to occur on a predictable basis, so that public sector users, like commercial users, could plan their spectrum transactions in advance. There might also have to be an "over-ride" system to deal with national security emergencies but this would be an exception rather than the rule.

There is a further dynamic which makes market participation by public sector spectrum users more palatable. In many countries, public sector bodies are perceived as having built up excessive spectrum holdings under the command-and-control period and in frequencies where there had been plenty of spectrum. In net terms, therefore, they are likely to be suppliers of spectrum to commercial users rather than net demanders. This is illustrated by the significant returns of spectrum to the regulator made in recent years by the French and UK Ministries of Defence. There may be exceptions – for example additional spectrum needed to provide emergency services with wireless broadband communications – but the basic flow of spectrum should be from public to private, by means of transactions made either by the spectrum users themselves or via the spectrum regulator.

This assimilation of public sector spectrum to the growing market place for commercial spectrum is precisely what the UK spectrum regulator, supported by the UK Government, is putting into practice. It was proposed in the Independent Audit noted above, and the policy of allowing public sector users to sell, lease, or share their spectrum, and of requiring them to acquire new spectrum in the market place, is now in the course of implementation in the UK [2].

If, however, public sector spectrum users are to participate firstly and effectively in spectrum markets, certain preconditions have to be fulfilled. In many jurisdictions, government departments, for example, are not issued with detailed licences specifying rights and responsibilities. If public sector spectrum is to become tradable, the associated property rights must be fully determined. Secondly, public sector spectrum use is, often for good reasons, scathed in secrecy. Whenever possible, information necessary for potential users to make decisions about the purchase, leasing or sharing of such frequencies has to be made available.

15.3 Intermediate steps to encourage efficiency in public sector spectrum use

Eliminating the boundary between private sector and public sector spectrum markets is a bold, if logical, step, but one that many spectrum regulators are as yet unwilling to take. For example, the European

Commission in its 2006 proposals for spectrum reform (see Chapter 6 above) advocates a market for much of the commercial spectrum but makes a broad exception for public service spectrum.

It is therefore useful to consider intermediate steps on the way to the application of market instruments; many of these are being considered as part of the US Government's steps towards reform.

(1) **Valuation of spectrum**. Public sector spectrum clearly has an opportunity cost, which is demonstrated either (in a market system) by the price realised by the auctioning or secondary trading of adjacent commercial spectrum, or can be computed by direct calculation of the opportunity cost, in the way described in Chapters 11 and 12 above.[1] The estimation of the value of spectrum used by each public sector user can crystallise and confirm intentions to interrogate those bodies vigorously as to whether their need for frequencies is fully justified.

(2) **Incorporation of valuations in investment decisions**. In all uses, spectrum costs represent only a part of the full costs of providing the service to end users. In most countries, the remaining costs, and sometimes the associated benefits, of a project such as a new weapons system are subject to some form of investment appraisal – for example of the project's lifetime costs. If a valuation of spectrum were available, it would be a simple matter to incorporate it as one of the costs of the project. This would encourage those engaged in procurement to examine the scope for substitution between spectrum and other costs. Such a requirement can be imposed in most countries by the Ministry of Finance or Treasury department, which supervises government spending.

(3) **Compensation of public sector users**. This is a method authorised by the United States Congress. Suppose some public sector (say, military) spectrum both has a valuable private use and can be replaced by alternative frequencies. The department

[1] The method described there yields an annual opportunity cost. This can be converted into a capital value by taking the sum of annual flows, each discounted by a cost of capital which reflects both general interest rate conditions in the economy, and the fairly high risks associated with spectrum assets, as a result of technological change which may change the pattern of spectrum use.

estimates the cost of transition to the new frequency, and the spectrum regulator then conducts an auction of current spectrum. If the auction proceeds exceed the "reserve price" represented by the transfer cost, the sale goes ahead, if otherwise, it does not, as the spectrum has been shown to be more efficiently employed in its current use. This can be seen as a form of "market testing". Care must be taken, however, to ensure that the costs of the transfer are not inflated by a public sector body seeking to get its hands on as much as possible of the auction proceeds.

(4) **Use of administrative prices**. A further major step is to impose a spectrum change, payable to the central government treasury or public sector uses. Ways of computing such charges were analysed in Chapters 10 and 11. We consider more fully in the next section how they can be made to stick in the public sector.

(5) **A market place with special pre-emption powers for the public sector**. The final step in the use of market methods is to allocate and assign public sector spectrum on the same footing as private sector frequencies. There may still be a need for arrangements to allow command-and-control methods to be used in an emergency to get spectrum for a vital and urgent purpose. The UK arrangements for pubic sector spectrum allocation contain such a provision. If it is not to undermine the discipline imposed by the market, there should be two conditions; first, the public sector users receiving the spectrum should pay a price at least equal to the market price; second, the procedure should be used sparingly and in strictly defined circumstances – otherwise users will not take responsibility for planning to meet their own needs.

This section has set out an incremental approach to improving efficiency in spectrum use by the public sector in the form of a "ladder" of options. Not all public sector users need to be on the same rung, but it is reasonable to expect the regime to get all users up at least as far as the incorporation of spectrum valuation in major procurement decisions.

However, the discussion so far has neglected one important feature of public sector organisations, the financial environment in which they operate. This must now be addressed.

15.4 Public sector incentive problems

An investor-owned operator selling services into a competitive mobile communications market will – it can confidently be predicted – seek to minimise its total costs. If it fails to do so, its costs may rise above those of its competitors, and it may go out of business.

Public sector spectrum users, by contrast, in many cases receive budgetary allocations to enable their activities to go ahead. Typically, the government, via the finance ministry, will set a target level of outputs, cost its provision, and make a budgetary allocation to cover the costs. If a department's costs are re-imbursed in this way it will have only a limited incentive to use inputs efficiently.

This applies to all inputs, including spectrum where there is a price for the input. If a department's budget were simply inflated to meet the cost of paying for spectrum, whatever it was, there would be no reason to economise on its use.

There are several ways in which this problem can be overcome, if the public spending regime allows it. For example, in some jurisdictions such as the UK, public spending plans are set three years in advance, based upon cost projections. If the costs turn out otherwise than projected, the department is responsible for making adjustments. Under this regime, a figure for expenditure on spectrum will be established in advance based upon an estimate of efficient usage. If the department incurs a lower cost, it can use the savings elsewhere. If it incurs a greater cost it must economise elsewhere. This can be supplemented by allowing departments to keep, perhaps up to a limit, the proceeds of their sale or leasing of surplus spectrum holdings.

Both of these assume a degree of discretion on the part of public sector organisations over their spending. In the United States for example, federal bodies may only spend in ways and in amounts authorised by the Congress, usually on an annual basis.

In such circumstances, economic incentives to encourage efficient spectrum use are less effective and they may have to be bolstered by more vigorous administrative efforts. These can involve monitoring and measuring spectrum use with rigorous interrogation of departments over the results, the mobilisation of "peer pressure" in application to poor performers, and in the limit removal of under-utilised spectrum. The obvious difficulty of this approach arises from asymmetry of information: on one hand, the public sector spectrum users are best informed about its real needs: on the other, the spectrum regulator knows that the user has an incentive to inflate its stated needs. If spectrum is removed on a random basis, the consequences can be very damaging. However, careful questioning by a knowledgeable auditor should disclose more flagrant misrepresentations.

15.5 Conclusions

This chapter has argued that there is no reason why public sector spectrum should be treated as "exceptional" and be immune from application of the market processes described elsewhere in this book. In practice, however, reforms in this area may have to take a more gradual course, relying on intermediate stages such as valuation and administrative prices.

There remains the pervasive problem, under almost any spectrum management regime, of generating incentives towards efficient use of spectrum by the public sector. In practice, there may be a continuing need for detailed scrutiny of public sector holdings and the application of pressure to hand back unneeded spectrum.

References

[1] See www.ntia.doc.gov/osmhome/spectrumreform/index.html.
[2] See www.spectrumaudit.org.uk.

16 Are developing countries different?

16.1 Introduction

Much of the discussion in previous chapters has revolved around problems of spectrum management likely to be encountered in developed countries. It is thus pertinent to ask how the situation differs for developing countries.

If anything, their dependence on spectrum-using technologies is even greater. Lacking fixed networks to deliver communications services such as voice telephony and broadcasting, they are heavily reliant on spectrum for commercial and non-commercial services. This is illustrated by recent ITU data, which show the growth of penetration (per 100 inhabitants) of fixed and mobile lines in the least developed countries. It shows that mobile lines were roughly the same in number as fixed lines in 2001 (see Figure 16.1), but by 2004 they outnumbered fixed lines by three to one. Over the 2000–4 period the number of television receivers, mostly relying on terrestrial distribution, also increased by 50%. These data emphasise the importance of getting spectrum policy right.

16.2 Consequences for spectrum management

What special aspects of spectrum management are important in developing countries? It is helpful first to identify the differentiating factors between the two environments that are relevant.

- Developing countries are characterised by a lower per capita income, which reduces consumption of all items including spectrum-using services.

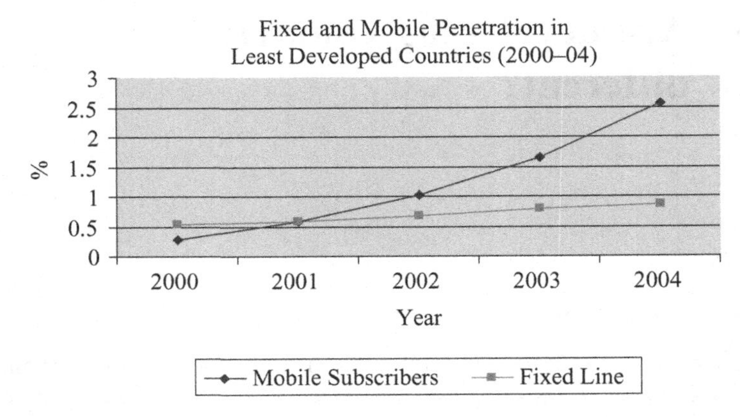

Figure 16.1. Relative penetration of fixed and mobile (based on ITU data).

- Conversely, a lack of alternative platforms places a high priority on the development of wireless systems; there is also growing evidence that mobile communications can improve business efficiency, widen markets and promote income growth in developing countries.
- At the same time, much spectrum in developing countries is not yet assigned, or assigned wastefully to government departments, especially defence forces.
- The combination of the above places a high priority on the speedy assignment of spectrum, much of which is historically under-utilised, to potential service providers. Some service operators are already illegally occupying frequencies, and opportunities exist for "normalising" their activities, if appropriate.
- In order to achieve the benefits of the expanding services, it is highly desirable that they are provided competitively: auctioning a single licensee to provide a monopoly service will not benefit end users.
- The assets of the spectrum regulator are limited in respect of finance, equipment and human resources.

This combination of factors places a premium on speed rather than precision in the award of spectrum licences. While the arguments in favour of the greater flexibility made possible by market methods apply in developing as well as in developed countries, if the design and implementation of a market mechanism causes delays, it may be better to issue licences initially on another basis – first come, first served, or a brief comparative hearing – if those methods can be protected from corruption and are implemented quickly.

However, the experience of Guatemala and San Salvador [1] with the introduction of a simple form of spectrum trading shows that a transition to market methods can be achieved without imposing substantial transactions costs. Certainly, these countries have problems of enforcement, especially in relation to illegal broadcasters, but the market regime has generated considerable flexibility in spectrum assignments.

Developing countries with relatively low incomes and sparse populations can also make greater use of unlicensed spectrum. Thus application of the decision tree set out in Section 14.3 above will probably generate many more decisions to create or expand commons than would be the case, for example, in Luxembourg or Los Angeles.

The resource problem can also be tackled by regional cooperation. This might occur, for example, among small Caribbean nations which might benefit from a coordinated policy on allocation and assignment of frequencies, as well as from the resources-saving which a combined approach might yield.[1]

An example of this is furnished by a project of the European Commission and the ITU to support the establishment of an integrated market for information and communications services in West Africa, which has a component dealing with spectrum management.

The guidelines for spectrum management which emerged from this project include the following.

- A common economic management regime for spectrum is desirable amongst the countries, and it will complement pro-competitive measures on telecommunications.
- A common framework for documenting and monitoring spectrum use is desirable, as this will help to combat interference. Regulators

[1] Note that in the European Union, a proposal has been made for a European Spectrum Agency to achieve similar goals.

should also develop a common template for a national table of frequency use.

- The objective of spectrum policy should be economic efficiency, and the achievement of public policy objectives.
- The countries should promote flexibility of spectrum use, while respecting international allocations.
- The spectrum management task should be undertaken by an independent regulatory agency also responsible for telecommunications.
- Spectrum, where unused, should be reassigned permanently or temporarily, or shared.
- All users should have an incentive to economise on spectrum use. Where frequencies have valuable potential use elsewhere, a charge for spectrum use should be made; otherwise spectrum charges should simply recover administrative costs.
- Restrictions in spectrum licences on spectrum use should be minimised.
- Spectrum trading should be considered.
- Auctions should be considered to assign spectrum licences in short supply [2].

16.3 Conclusion

We have concluded that the goals of spectrum management are broadly the same in developing as in developed countries, as are the available instruments. However, their governments and regulators may face less congestion and have an incentive to speed up the assignment process in order to provide services to their citizens. This may imply either the development of simple-to-administer market processes or, in the interim, fast track administrative processes.

References

[1] T. Hazlett and W. Leighton, “Property rights to radio spectrum in Guatemala and El Salvador”, 2006. Available at hermes.ssrn.com/sol3/papers.cfm?abstract_id=889409.

[2] C. Doyle and M. Schaar (2005), *Harmonization of Policies Governing the ICT Market in the UEMOA-ECOWAS Space: Radio Spectrum Management*. ITU; available at http://www.itu.int/ITU-D/treg/Events/ Seminars/ITU-EC-Project /Ghana /modules/ FinalDocuments/Spectrum.pdf.

IV Conclusions

17 Conclusions

17.1 A reminder of the problem

At the start of this book we said that spectrum needs to be managed to avoid the interference between different users becoming excessive. We noted that the key purpose of spectrum management is to maximise the value that society gains from the radio spectrum by allowing as many users as possible while ensuring that the interference between different users remains manageable. Then we observed that the current "command-and-control" approach was unlikely to achieve this objective and was becoming more difficult to manage as an ever expanding range of applications arose. Instead, we noted how increasingly spectrum managers were turning to economic management methods to achieve their duties.

17.2 Key conclusions

We made the following conclusions.

- **Technological advances**. The advance of technology is having some impact on spectrum management. Multi-modal radios are gradually reducing the advantages of international harmonisation, making it easier for regulators to allow the use of market forces. Technologies that provide "underlays" (UWB) or "overlays" (cognitive radio) might require radical changes to spectrum management, but in practice cognitive radio may be best enabled simply by providing spectrum owners with sub-leasing capabilities while UWB can be accommodated as an increased noise floor for existing owners.
- **Division of spectrum**. While there are many alternative methods of dividing access to spectrum, all can be accommodated as a subdivision within the current overall process of division by

frequency. For example, after a division by frequency, it is possible to further subdivide by time, angle, polarisation, geography or use. The regulator could choose to make this division themselves, or they could provide the licence holder with the freedom to do so, perhaps through the flexible type of licence envisaged under trading. Hence, major changes to spectrum management caused by technology look unlikely.

- **Market forces**. The key tools that a regulator needs to deploy in order to allow market forces to manage spectrum are auctions, trading and property rights. In addition, some powers to address anti-competitive behaviour may also be required.
- **Auctions**. Spectrum auctions have been used extensively around the world for assigning many thousands of licences covering many different uses and types of user. Since the first spectrum auction in 1989, the design of auctions has evolved to accommodate very large scale auctions and small scale auctions. The simultaneous ascending auction format is often used for large scale auctions and is particularly good where licences may be similar though not identical and there are complementarities. For small scale auctions where there is little package bidding, sealed bid auctions would seem the most appropriate. More recently auction theorists have been debating the merits of more sophisticated auction formats that combine the benefits of both sealed bid and ascending bid auction formats such as the Anglo-Dutch auction and the clock-proxy auction.
- **Trading**. Trading is likely to work well where no single player has market power, property rights are well defined, full information about previous trades and the current trade is available and there are no unforeseen externalities. Possible issues with trading include a lack of harmonisation and windfall profits, but we do not believe these are serious. Trading is a valuable tool for many parts of the radio spectrum but not all. Radio spectrum is not homogeneous – different frequencies can support different types of applications – and this means that there are potentially many different markets for spectrum, depending on frequency, usage and users. Regulatory enforcement of property rights may be required on a greater scale in some markets than others. Where the costs of regulation would

outweigh the benefits, there may be more effective ways of managing the radio spectrum. For example, for some uses and frequency ranges it might be more effective to make radio spectrum a public good (i.e. licence exempt spectrum).

- **Property rights**. A set of property rights is critical to facilitate trading and enable change of use. The key technical components of property rights comprise limits on in-band, out-of-band and geographical emissions. From these limits, neighbours will be able to calculate their likely interference levels. These rights and obligations will need to be carefully defined since the interference experienced by one user depends on the rights granted to other users. Within these rights easements should not usually be allowed, but rights should be allocated in ways which take account of the economic value, and interference potential, of new technologies such as UWB. A compulsory purchase power for spectrum should be confined to national security needs. Licences should be perpetual but spectrum licensees should not pay a perpetual annual charge or any charge which discourages efficient trading.
- **Competition issues**. Concentration of spectrum holdings in the hands of one or a small number of operators in commercial markets can provide a means to monopolise service markets. The extent of the risk depends on policies adopted at the time of a spectrum award and the rules governing spectrum holdings thereafter. There are grounds for designing award rules in ways which will prevent any operator from immediately gaining market power in a services market. This applies to both command-and-control and market-based spectrum management regimes. Spectrum markets can provide a degree of flexibility and responsiveness to changing market conditions which is beyond the reach of command-and-control methods. Moreover, intervention by the spectrum regulator or by a competition or communications services regulator can reduce the risk of market failure associated with abuse of power in spectrum markets. For these reasons, concerns about competition are not a valid basis for rejecting spectrum markets. Sufficient instruments exist to combat monopolisation of spectrum in a market context, and

the alternative command-and-control methods can have equally deleterious effects on end users.

- **Band managers**. There are many potential types of commercial band manager including site owners, profit making organisations offering licensed access and profit making organisations offering unlicensed access. Under an environment where spectrum is tradable, the key to successful operation will be the ability to make use of the spectrum more efficiently than the regulator. However, this is difficult to do, making the economics of band management problematic. If regulators wish to encourage band managers they may need to look into specific policy measures to do so. More likely, they will adopt a neutral view and allow the most efficient forms of arrangements to result through the operation of the market.
- **Spectrum pricing**. We have shown that a spectrum management agency can use prices to achieve efficiency, and that generally this will lead to superior outcomes than pricing on the basis of cost recovery. We discussed the economic underpinnings of efficient pricing of spectrum and illustrated how prices can be devised. The Smith–NERA methodology was discussed at length, and we indicated how this algorithm can be deployed in practice. We also discussed how spectrum pricing can coexist alongside trading. Charging annual fees for the holding of spectrum is one way in which the spectrum manager can encourage current and prospective holders to make the right decisions to ensure efficient use of the spectrum. We noted that many holders of spectrum are not in a position to make rapid changes to their use of spectrum in response to the application of AIP, but that in practically every case the holders of spectrum have opportunities to change their use of spectrum in the longer term. The use of AIP is justified by the benefits that should materialise in the longer term, as better decisions are made in light of increased awareness and appreciation of the value of spectrum – better decisions that should lead to more efficient use of the spectrum.
- **Commons**. An economic analysis of the situation suggests that spectrum should be unlicensed where there is little probability of congestion. Despite arguments about the ability of "spectrum

commons" to alleviate congestion, congestion across key parts of the spectrum is likely for the foreseeable future. Congestion is unlikely where short range communications are used and can be made less likely by regulatory insistence on, for example, politeness protocols. Hence, there should be a mix of licensed and unlicensed spectrum with the unlicensed approach restricted to bands and applications where congestion is unlikely. Market mechanisms cannot be used in determining how much unlicensed spectrum would be needed so instead the regulator needs to make a judgement. We set out two approaches that the regulator could follow. The top-down approach estimated the total amount of unlicensed spectrum that might be needed and attempted to find sufficient spectrum to meet this need. The bottom-up approach considered bands of spectrum on a band-by-band basis, using a structured process to determine for each band whether licensed or unlicensed use would be most appropriate.

- **Public spectrum**. There is no reason why public sector spectrum should be treated as "exceptional" and be immune from application of the market processes. In practice, however, reforms in this area may have to take a more gradual course, relying on intermediate stages such as valuation and administrative prices. There remains the pervasive problem, under almost any spectrum management regime, of generating incentives towards efficient use of spectrum by the public sector. In practice, there may be a continuing need for detailed scrutiny of public sector holdings coupled with the application of pressure to hand back unneeded spectrum.
- **Developing countries**. The goals of spectrum management are broadly the same in developing as in developed countries, and so are the available instruments. However, their governments and regulators may face less congestion and have an incentive to speed up the assignment process in order to buy services for their citizens. This may imply either the use of fast-track administrative methods or the development of a simple to administer market process.

17.3 In summary

Optimally managing the radio spectrum can no longer rely on command-and-control methods. Instead, economic tools need to be used. In this book we have examined the tools that would be needed, discussed how they might be formulated and deployed, and considered the key objections to their use. We have shown that, while far from simple, it is possible to devise an economic approach to the management of radio spectrum which is more likely to optimise the use of the spectrum than the existing approach. We expect to see this approach increasingly deployed by regulators around the world over the coming years.

Further reading

Chapters 2 and 3

M. Schwartz, *Mobile Wireless Communications*, Cambridge University Press, 2005.

K. Siwiak and D. McKeown, *Ultra-wideband Radio Technology*, Wiley, 2004.

W. Webb, *Wireless Communications: The Future*, Wiley, 2007.

Chapter 4

M. Cave, "Review of Radio Spectrum Management: an independent review for Department of Trade and Industry and HM Treasury", March 2002, available at http://www.ofcom.org.uk/static/archive/ra/spectrum-review/index.htm

"The FCC report to congress on spectrum auctions", Federal Communications Commission, Wireless Telecommunications Bureau WT Docket No. 97–150, 30 September 1997, available at http://wireless.fcc.gov/auctions/data/papersAndStudies/fc970353.pdf

Chapter 5

There is a large literature on the theory and practice of spectrum auctions. The following are very good accounts of contemporary spectrum auction analysis:

P. Cramton, "Spectrum Auctions", in *Handbook of Telecommunications Economics*, M. Cave, S. Majumdar and I. Vogelsang (eds)., Amsterdam: Elsevier Science B.V., 2002, chapter 14, pp. 605–639.

P. Milgrom, *Putting Auction Theory to Work*, Cambridge University Press, 2004.

P. Klemperer, *Auctions: Theory and Practice*, Princeton University Press, 2004.

The following report offers a detailed insight into the 3G spectrum auction held in the UK in 2000:

"The Auction of Radio Spectrum for the Third Generation of Mobile Telephones", report by the comptroller and auditor general, HC 233, Session 2001–2002: 19 October 2001. Available at http://www.nao.org.uk/publications/nao_reports/01–02/0102233.pdf.

The following article is a seminal contribution and an excellent read:

W. Vickrey, "Counterspeculation, auctions, and competitive sealed tenders", *Journal of Finance*, **16**, 8–37, 1961.

Chapter 6

The seminal contribution is:

R. Coase, "The Federal Communications Commission", *Journal of Law and Economics*, **II**, 1–40, 1959.

See also

R. Coase, "The problem of social cost", *Journal of Law and Economics*, **3**, 1–44, 1960.

A comprehensive report on secondary trading of radio spectrum was undertaken for the European Commission and published in 2004:

Analysys, DotEcon and Hogan & Hartson, "Study on conditions and options in introducing secondary trading of radio spectrum in the European Community", final report for the European Commission, May 2002. Available at http://europa.eu.int/information_society/policy/radio_spectrum/docs/ref_docs/secontrad_study/secontrad_final.pdf

M. Cave, "Review of Radio Spectrum Management: an independent review for Department of Trade and Industry and HM Treasury", March 2002.

T. Hazlett, "Assigning property rights to radio spectrum users: why did FCC license auctions take 67 years?", *Journal of Law and Economics*, **41** (2), 529–575, 1998.

Ofcom, "A Statement on Spectrum Trading Implementation in 2004 and beyond", August 2004.

P. Spiller and C. Cardilli, "Towards a Property Rights approach to Communications Spectrum", *Yale Journal on Regulation*, **16**, 53–83, 1999.

T. Valletti, "Spectrum Trading", *Telecommunications Policy*, **25**, 655–670, 2001.

L. White, "'Propertyzing' the Electromagnetic Spectrum: Why It's Important, and How to Begin", in J. A. Eisenach and R. J. May (eds.), *Communications, Deregulation and FCC Reform: Finishing the Job*, Kluwer, 2001.

Chapter 7

See http://www.ofcom.org.uk/consult/condocs/sur/, *where a range of information including a consultation and a detailed report on Spectrum Usage Rights can be found.*

Chapter 8

R. Coase, "The Federal Communications Commission", *Journal of Law and Economics*, 1959.

R. Coase, "The problem of social cost", *Journal of Law and Economics*, **3**, 1–44, 1960.

T. Hazlett, "The spectrum allocation debate: an analysis", *IEEE Internet Computing*, Sept/Oct 2006, pp. 52–58.

G. Faulhaber, "The future of wireless telecommunications: spectrum as a critical resource", *Information Economics and Policy*, 2006.

Chapter 10

T. Hazlett, "The Wireless Craze, The Unlimited Bandwidth Myth, The Spectrum Auction Faux Pas, and the Punchline to Ronald Coase's 'Big Joke': An Essay on Airwave Allocation Policy", *Harvard Journal of Law and Technology*, Spring 2001.

Chapters 11 and 12

"An Economic Study To Review Spectrum Pricing", Indepen, Aegis Systems and Warwick Business School, February 2004, available at http://www.ofcom.org.uk/research/radiocomms/reports/independent_review/spectrum_pricing.pdf.

Chapters 13 and 14

See http://www.ofcom.org.uk/consult/condocs/sfr/ where the Ofcom Spectrum Framework Review and related material can be found.

Y. Benkler, "The Wealth of Networks: How Social Production Transforms Markets and Freedom", *Yale University Press,* April 2006. (A complete PDF of the book is freely downloadable on the wiki of the book and is available under a *Creative Commons Noncommercial Sharealike license.)*

Y. Benkler, "Some Economics of Wireless Communications", *16 Harvard Journal of Law & Technology*, **25** (Fall 2002).

Kevin Werbach, "Open Spectrum: The New Wireless Paradigm", http://werbach.com/docs/new_wireless_paradigm.htm.

Chapter 15

M. Cave, "Independent review of major spectrum holding", HMSO, 2005.

Chapter 16

B. Wellenius and I. Neto, "The radio spectrum: opportunities and challenges for the developing world", *INFO*, **2**, 18–33, 2006.

INFODEV/ITU ICT Regulation Toolkit, *Module 5: Radio Spectrum management* available at http://www.ictregulationtoolkit.org/.

Abbreviations

AIP	administrative incentive pricing
AM	amplitude modulation
AWS	advanced wireless services
BFWA	broadband fixed wireless access
CDMA	Code Division Multiple access
CEPT	Confederation of European Post and Telecommunication Organisations
CMRS	commercial mobile radio services
DFS	dynamic frequency selection
DSP	digital signal processor
EC	European Commission
FCC	Federal Communications Commission
FM	frequency modulation
ITU	International Telecommunications Union
JRC	joint radio company
MPL	minimum path length
MVNO	mobile virtual network operator
MW	medium wave
NB	net benefit
NTIA	National Telecommunications and Information Administration
OOB	out-of-band
PAMR	public access mobile radio
PBR	private business radio
PFD	power flux density
QoS	quality of service
RA	Radiocommunications Agency
RFID	radio frequency identification

SAA	simultaneous ascending auction
SAMR	simultaneous ascending multiple round
SDR	software defined radio
STU	standard trading unit
SW	short wave
TDD	time division duplex
UWB	ultra-wideband
VHF	very high frequency

Author biographies

Martin Cave

Martin Cave is Professor at Warwick Business School and director of its Centre for Management under Regulation. His academic work is in regulatory economics, especially of the telecommunications. He is co-editor of the *Handbook of Telecommunications Economics*, Vol. 1 (2002), Vol. 2 (2005) and *Digital Broadcasting* (2006), co-author of *Understanding Regulation* (1999), and author of numerous academic articles. In the past five years Martin has completed two major reports for the UK Government – *the Review of Spectrum Management* (2002) and *the Audit of Major Spectrum Holdings* (2005). The former set out an overall strategy for introducing market forces into frequency management; the latter proposed ways of promoting efficiency in spectrum use in the public sector. The recommendations of both reports were accepted and implemented by the UK Government. Martin is also co-author of a web-based guide to spectrum management prepared under the auspices of the World Bank (InfoDev) and the ITU.

Martin has also advised many regulatory bodies in Europe, Asia and Australasia, and in 2006 he was special advisor on regulatory reform to the European Commissioner for the Information Society and Broadcasting.

Chris Doyle

Chris Doyle is a Senior Research Fellow in the Centre for Management under Regulation, Warwick Business School. He is an economist working on competition and regulatory issues of relevance to network industries and telecommunications. He has authored many academic articles and contributed to several books. He was Chairman of the

ITU's 3G Licensing Workshop in 2001, and in 2005 wrote the *Radio Spectrum Management Report* commissioned by the ITU and European Commission for the West African Telecommunications Regulatory Assembly. He has written reports on spectrum pricing for the Dutch Ministry of Finance and Ofcom in the UK. He designed the successful Nigerian GSM auction in 2001 and has advised a number of companies on bidding strategy in spectrum auctions. He serves as an advisor to a number of governments on telecommunications policy.

William Webb

William joined Ofcom as Head of Research and Development just prior to it officially becoming the national telecoms regulator. Key outputs from the team include the Annual Technology Review and a range of research reports. William also leads across a wide range of spectrum strategy. He wrote the Spectrum Framework Review – Ofcom's long term framework for radio spectrum management and led on ultra-wideband strategy and property rights for spectrum ("spectrum usage rights"). He is a member of Ofcom's Senior Management Group.

William has worked in the wireless communications industry since his graduation in 1989. He has worked on many research and development projects including detailed research into QAM for mobile radio, directing multiple propagation measurements, designing and writing a microcell propagation tool and producing hardware demonstrators for early DECT and QAM systems. He took the lead technology role on the "Smith–NERA" studies into the economic value of spectrum and spectrum pricing and played a key part in the standardisation of the European Railway's GSM technology.

He is a judge for the Wall Street Journal's Annual Innovation Award, a Visiting Professor at the Centre for Communications Systems Research at Surrey University and a Vice President of the IET. He is a Fellow of the Royal Academy of Engineering and the IET. He has published over 70 papers in a mix of publications spanning learned journals to the Wall Street Journal. His books include *Modern Quadrature Amplitude Modulation* published by John Wiley and the IEEE in

1994, *Introduction to Wireless Local Loop* published by Artech House in 1998, *Understanding Cellular Radio* published by Artech House in 1998, *The Complete Wireless Communications Professional* published by Artech House in 1999, *Single and Multi-carrier QAM* published by Wiley in 2000, *Introduction to Wireless Local Loop – second edition: Broadband and Narrowband Systems* published by Artech House in 2000, and *The Future of Wireless Communications* published by Artech House in 2001. He is a series editor for this book series. His biography is included in Debrett's *People of Today* and multiple *Who's Who* publications around the world.

Subject index